Har

KU-370-912

A SHORT COURSE IN CLOUD PHYSICS

SECOND EDITION

INTERNATIONAL SERIES IN NATURAL PHILOSOPHY
VOLUME 96
GENERAL EDITOR: D. TER HAAR
A List of other Titles on Natural Philosophy follows Index

SOME OTHER TITLES OF INTEREST

BRUCE, J. P. & CLARK, R. H.
Introduction to Hydrometeorology

GRIFFITHS, D. H. & KING, R. F.
Applied Geophysics for Engineers and Geologists

McLELLAN, H. J.
Elements of Physical Oceanography

PATERSON, W. S. B.
The Physics of Glaciers

PICKARD, G. L.
Descriptive Physical Oceanography

SCORER, R. S. & WEXLER, H. A.
A Colour Guide to Clouds

A SHORT COURSE IN CLOUD PHYSICS

by

R. R. ROGERS
Professor of Meteorology, McGill University

SECOND EDITION

PERGAMON PRESS
OXFORD · NEW YORK · TORONTO · SYDNEY
PARIS · FRANKFURT

U.K.	Pergamon Press Ltd., Headington Hill Hall, Oxford OX3 0BW, England
U.S.A.	Pergamon Press Inc., Maxwell House, Fairview Park, Elmsford, New York 10523, U.S.A.
CANADA	Pergamon of Canada, Suite 104, 150 Consumers Road, Willowdale, Ontario, M2 J1P9 Canada
AUSTRALIA	Pergamon Press (Aust.) Pty. Ltd., P.O. Box 544, Potts Point, N.S.W 2011, Australia
FRANCE	Pergamon Press SARL, 24 rue des Ecoles, 75240 Paris, Cedex 05, France
FEDERAL REPUBLIC OF GERMANY	Pergamon Press GmbH, 6242 Kronberg-Taunus, Pferdstrasse 1, Federal Republic of Germany

First edition 1976

Second edition 1979

Reprinted 1979

British Library Cataloguing in Publication Data

Rogers, Roddy Rhodes
A short course in cloud physics.— 2nd
ed.—(International series in natural
philosophy).—(Pergamon international
library).
1. Cloud physics
I. Title   II. Series
551.5'76   QC921.5   78-40104

ISBN 0-08-023040-7 (Hardcover)
ISBN 0-08-023041-5 (Flexicover)

*Printed in Great Britain by
Biddles Ltd, Guildford, Surrey*

CONTENTS

PREFACE

THIS book is based on lecture notes for the thermodynamics and cloud physics parts of an introductory course in physical meteorology. It is designed for graduate students with little or no previous exposure to meteorology. Most of the material is also suitable for upper-level undergraduate students in science or engineering.

The first four chapters cover topics in atmospheric thermodynamics that are essential for understanding cloud physics. The central chapters are on the formation of cloud droplets, growth by condensation and collection, and the development of rain and snow. Comprising the heart of traditional cloud physics, these topics are treated extensively in the books by Fletcher (1962), Byers (1965), and Mason (1971), whose influence on this textbook is gratefully acknowledged. The later chapters are devoted mainly to observations of precipitation and the way these relate to microphysical theory. Principles of weather radar are outlined and a brief account is given of weather modification. The final chapter introduces the numerical modeling of clouds, a subject of considerable current interest.

In this edition some of the errors in the earlier one have been corrected, a few derivations have been recast, and a small amount of new material has been added. Answers to alternate problems have been included, in response to the suggestion of several teachers who used the earlier edition.

For their comments and suggestions, I wish to thank those who wrote to me about the earlier edition. I am especially grateful to David Robertson, Henry G. Leighton, and Isztar I. Zawadzki for their advice and help.

Montreal, October 1977   R. R. ROGERS

INTRODUCTION

ONE of the branches of physical meteorology, cloud physics may be defined as the science of clouds in the atmosphere. This broad definition admits a great body of knowledge, ranging for example from the classification of clouds to the chemistry of rainwater. It is generally agreed, however, that the principal task of cloud physics is to explain the formation of clouds and the development of precipitation. Accordingly, this book is concerned with cloud and precipitation development at the expense of other important parts of cloud physics, such as radiative transfer in clouds and optical and electrical phenomena.

Processes of two very different kinds are responsible for cloud and precipitation development. First, it is required that regions or patches of air become essentially saturated, that is, develop relative humidities close to 100%. This is accomplished almost entirely by vertical motions in cloud-free air. These vertical motions have horizontal extents ranging from tens of meters to hundreds of kilometers, depending upon how they are produced. In magnitude the vertical velocities range from centimeters per second to tens of meters per second, depending also on the generating process. Such vertical motions are required for cloud formation and play a dominant role in determining the character and quantity of precipitation that the cloud ultimately produces. These relatively large-scale processes—all involving air motions—are generally referred to as "cloud dynamics" or "cloud kinematics".

Cloud processes of the second kind are on a much smaller scale—a scale comparable in size to the dimensions of individual cloud and precipitation particles. These are the processes of cloud droplet formation, growth, and interplay with environment. Involved are elements of thermodynamics, diffusion theory, and physical chemistry. The goal of this phase of cloud physics is to explain the circumstances by which an individual cloud droplet can form from the vapor phase, grow to visible size, and interact with other cloud particles to form precipitation. These aspects of cloud physics are combined under the term "cloud microphysics".

Books on cloud physics usually emphasize microphysics. Dynamic processes are no less important, but are less well understood than the microphysics. There are several reasons for this. While many microphysical processes can be studied in the laboratory, the large-scale, dynamic effects cannot so readily be duplicated. Also there are serious unsolved theoretical problems in cloud dynamics, whereas the theoretical basis of microphysical processes is more complete. Finally, adequate observations of the airflow in clouds have become available only in the last decade, with the application of Doppler radar to cloud studies.

Following tradition, this book is devoted mainly to microphysics. Emphasis is on the development of precipitation, however, with the traditional topics of nucleation, aerosols, and instrumentation receiving only brief treatment. The first four chapters cover elements of atmospheric thermodynamics relevant to the study of clouds. These are followed by chapters on the microphysics of clouds and precipitation. An outline of the principles of weather radar provides a background for the observational material in later chapters on precipitation processes and severe storms. The final chapter introduces the topic of numerical cloud modeling and will lead the interested reader to the relevant references. Here the complex interactions of dynamics and microphysics come into focus and an indication is given of the extent to which theory, with its necessary approximations, is able to explain the observations.

CHAPTER 1

THERMODYNAMICS OF DRY AIR

Atmospheric composition

Air is a mixture of several so-called permanent gases, a group of gases with variable concentrations, and different solid and liquid particles of variable concentrations. Nitrogen and oxygen account respectively for about 78% and 21% by volume of the atmosphere's permanent gases, with the remaining 1% consisting primarily of argon, but with trace amounts of neon, helium, and other gases. The composition of air is remarkably uniform, with the relative proportions of these permanent gases being essentially the same the world over and up to an altitude of 90 km.

The most abundant of the gases present in variable amounts are water vapor, carbon dioxide, and ozone. These are the gases that strongly affect radiative transfer in the atmosphere. Water vapor also plays a central role in atmospheric thermodynamics.

The particles of solid and liquid material suspended in the air are called aerosols. Common examples are the water droplets comprising clouds, ordinary dust particles, and pollen. Thermodynamics is concerned with the gases, but a select group of the aerosols called hygroscopic nuclei are crucial for the condensation of water in the atmosphere.

The approach in meteorology is to treat air as a mixture of two ideal gases: "dry air" and water vapor. This mixture is called moist air. The thermodynamic properties of moist air are determined by combining the separate thermodynamic behaviors of dry air and water vapor.

Equation of state for dry air

The equation of state for a perfect gas, or ideal gas law, expresses the relationship between pressure p, volume V, and temperature T

1

of a gas in thermal equilibrium:

$$pV = CT, \tag{1.1}$$

with C a constant depending upon the particular gas.

The equation is reduced to standard form by employing Avogadro's Law, which states that at the same pressure and temperature, the volume occupied by one mole of any gas is the same. Denoting this volume by v, we have

$$pv = C'T, \tag{1.2}$$

where now C' is the same constant for all gases. It is called the *universal* gas constant and denoted by R^*. $R^* = 8.314 \times 10^7$ ergs K^{-1} mole^{-1} = 8.314 joules K^{-1} mole^{-1}.

Since an arbitrary volume $V = nv$, with n the number of moles, it follows from (1.2) that

$$pV = nR^*T. \tag{1.3}$$

Dividing by the mass M of the gas gives

$$\frac{pV}{M} = \frac{n}{M} R^*T.$$

But $V/M = \alpha$, the specific volume, and $n/M = 1/m$ where m denotes the molecular weight of the gas. Consequently (1.3) reduces to

$$p\alpha = R'T, \tag{1.4}$$

where $R' = R^*/m$ is called the *individual* gas constant.

It is possible to calculate the effective molecular weight of dry air by suitably averaging the molecular weights of nitrogen, oxygen, etc., of which it is composed. This turns out to be 28.9 g/mole. Accordingly, the individual gas constant for dry air is

$$R' = 2.87 \times 10^6 \text{ ergs g}^{-1} K^{-1} = 287 \text{ J kg}^{-1} K^{-1}.$$

Over the meteorological range of temperature and pressure, (1.4) describes the behavior of dry air with sufficient accuracy for most purposes.

The first law of thermodynamics

The first law is a statement of two empirical facts:

1. Heat is a form of energy.
2. Energy is conserved.

The first of these is called Joule's Law, and expresses the mechanical equivalent of heat as

$$1 \text{ cal} = 4.186 \times 10^7 \text{ ergs} = 4.186 \text{ J.} \qquad (1.5)$$

The second of these empirical facts may be stated in algebraic form:

$$dQ = dU + dW. \qquad (1.6)$$

Of the total amount of heat added to a gas, dQ, some may tend to increase the internal energy of the gas by amount dU, and the remainder will cause work to be done by the gas in the amount dW. It is generally more useful to express this relation for a unit mass of gas, in which case (1.6) becomes

$$dq = du + dw. \qquad (1.7)$$

We examine first the work term in (1.7). Consider a parcel of gas with volume V and surface area A, as illustrated in Fig. 1.1. The change in volume associated with a small incremental linear expansion dn is

$$dV = A\,dn.$$

But $p = F/A$, where F is the force exerted by the gas, so that

$$p\,dV = F\,dn. \qquad (1.8)$$

The work done by the gas in expanding is $dW = F\,dn$. Consequently

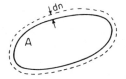

FIG. 1.1. Expanding parcel of gas.

(1.8) may be written

$$dW = pdV.$$

The work done per unit mass of gas (specific work) is

$$dw = pd\alpha. \tag{1.9}$$

In general, the specific work done in a finite expansion from α_1 to α_2 is

$$\int dw = \int_{\alpha_1}^{\alpha_2} pd\alpha.$$

This integration may be visualized with the help of a thermodynamic diagram.

A thermodynamic diagram is a chart whose coordinates are variables of state. A given equilibrium thermodynamic state of a gas may be represented by a point on such a chart. As a gas goes through successive equilibrium states (for example in response to heating or to an external force), it traces out a path on a thermodynamic diagram.

The work done by a gas in expanding is readily illustrated on a chart with coordinates of pressure versus specific volume, as in Fig. 1.2. In the example shown the gas expands from initial state $A(p_1,\alpha_1)$ to final state $B(p_2,\alpha_2)$. The specific work done is represented by the area $ABCD$. There are actually any number of possible equilibrium paths from A to B, depending upon whether heat is added to or taken from the gas, and at what point during the process this heat transfer occurs. The work done depends on the path of integration, which is another way of saying $dw = pd\alpha$ is not

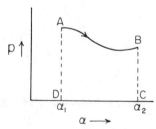

FIG. 1.2. Thermodynamic chart representing work done in expansion.

an exact differential. Thus, there does *not* exist a general function $F(\alpha)$ such that

$$\int_{\alpha_1}^{\alpha_2} p\,d\alpha = \int_{\alpha_1}^{\alpha_2} dF(\alpha) = F(\alpha_2) - F(\alpha_1).$$

(Some authors are careful to indicate this fact by writing $đw$ instead of dw.)

Of special interest in thermodynamic theory are *cyclic* processes, in which the gas undergoes a continuous series of changes in state, but ends up with the same thermodynamic coordinates it had initially. One such cyclic process is pictured in Fig. 1.3 on a

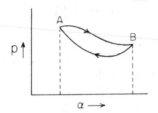

FIG. 1.3. Cyclic process.

p,α-diagram. The gas starts at state B and proceeds to state B along the indicated curve. As before, the area under this curve gives the work done by the gas in expanding from α_1 to α_2. Next the gas is compressed and made to return to state A along the lower curve. In this step of the process work is done *on* the gas. The net work done by the gas in the complete cyclic process is given by the hatched area. Note that if the process had taken place in the opposite sense, with arrows reversed, the hatched area would stand for net work done *on* the gas. The net work in a cyclic process is described mathematically by an integral over the path,

$$\int_c dw = \int_c p\,d\alpha,$$

where the subscript c denotes integration about a closed path. For integrands that are exact differentials, any such cyclic integration yields zero, because the integral over an exact differential depends only on the limits of integration. This is not the case in general for dw, as explained above.

Next we consider the du term in (1.7). For an ideal gas, any increase in internal energy appears as an increase in temperature. The temperature change is proportional to the amount of heat added according to

$$dT = \frac{1}{c} dq, \tag{1.10}$$

where c is called the specific heat capacity and is measured in cal $g^{-1}\,°C^{-1}$. For a gas, c is not constant but depends upon whether work is done while the heat is added. If no work is done, $d\alpha = 0$ from (1.9) and for the specific heat we write

$$c_v = \left(\frac{dq}{dT}\right)_\alpha, \tag{1.11}$$

known as the specific heat at constant volume.

Another case of interest is the addition of heat with pressure held constant, for which process the specific heat is given as

$$c_p = \left(\frac{dq}{dT}\right)_p, \tag{1.12}$$

and called the specific heat at constant pressure.

Evidently $c_p > c_v$, since in a constant pressure process some of the added heat will be used in the work term $pd\alpha$, while in the constant volume process all added heat goes toward increasing T. For dry air,

$$c_p = 0.240 \text{ cal g}^{-1}\,K^{-1} = 1004 \text{ J kg}^{-1}\,K^{-1}$$
$$c_v = 0.171 \text{ cal g}^{-1}\,K^{-1} = 717 \text{ J kg}^{-1}\,K^{-1}$$

Of the total heat added, the amount that goes into the internal energy is

$$du = c_v dT \tag{1.13}$$

and the remainder goes into the work term. Thus the general expression for the conservation of energy is

$$dq = c_v dT + pd\alpha. \tag{1.14}$$

Special processes

By differentiating (1.4), we obtain

$$pd\alpha + \alpha dp = R'dT \qquad (1.15)$$

as a differential equation relating changes of pressure, specific volume, and temperature under conditions of thermodynamic equilibrium. Combining (1.15) and (1.14),

$$dq = (c_v + R')dT - \alpha dp.$$

But

$$c_p = \left(\frac{dq}{dT}\right)_p = c_v + R',$$

so that

$$dq = c_p dT - \alpha dp \qquad (1.16)$$

may be used instead of (1.14) as an alternate expression of the first law.

Certain special processes are now defined in terms of these equations.

(a) Isobaric process: $dp = 0$

$$dq = c_p dT = \left(\frac{c_p}{c_v}\right) c_v dT = \left(\frac{c_p}{c_v}\right) du. \qquad (1.17)$$

(b) Isothermal process: $dT = 0$

$$dq = -\alpha dp = pd\alpha = dw. \qquad (1.18)$$

(c) Isosteric process: $d\alpha = 0$

$$dq = c_v dT = du. \qquad (1.19)$$

(d) Adiabatic process: $dq = 0$

$$c_p dT = \alpha dp \qquad (1.20)$$

or

$$c_v dT = -pd\alpha. \qquad (1.21)$$

The adiabatic process is of special significance because many of the temperature changes that take place in the atmosphere can be approximated as adiabatic. From (1.20) and the equation of state,

$$c_p dT = R'T\frac{dp}{p}, \qquad (1.22)$$

which may be integrated to give

$$\left(\frac{T}{T_0}\right) = \left(\frac{p}{p_0}\right)^k,$$ (1.23)

where $k = R'/c_p = (c_p - c_v)/c_p = 0.286$.

The result (1.23) is called Poisson's equation for adiabatic processes. It is possible to derive expressions equivalent to (1.23) relating any two of the thermodynamic variables pressure, temperature, and specific volume.

A fourth thermodynamic variable, called the potential temperature, is defined on the basis of (1.23). It is denoted by θ and defined by

$$\left(\frac{T}{\theta}\right) = \left(\frac{p}{1000 \text{ mb}}\right)^k$$

or

$$\theta = T\left(\frac{1000 \text{ mb}}{p}\right)^k$$ (1.24)

and may be interpreted as the temperature that a parcel of air would have if, starting with temperature T at pressure p, it were subjected to an adiabatic compression or expansion to a final pressure of 1000 mb. (1 mb $= 10^{-3}$ bars $= 10^3$ dynes/cm^2 $= 10^2$ Pa.) The potential temperature is called a variable of state, because it is expressible in terms of the state variables p and T. In any adiabatic process, θ is a constant. We say that potential temperature is a *conservative* property with respect to adiabatic processes.

Entropy

The second law of thermodynamics implies the existence of another variable of state, termed entropy, which may be defined by the equation

$$d\phi = \frac{dq}{T},$$ (1.25)

where $d\phi$ is the increase in (specific) entropy accompanying the addition of heat dq to a unit mass of gas at temperature T. It follows from (1.16) that

$$d\phi = \frac{1}{T}[c_p dT - \alpha dp] = c_p \frac{dT}{T} - R' \frac{dp}{p} = c_p\left[\frac{dT}{T} - k\frac{dp}{p}\right] = c_p \frac{d\theta}{\theta}.$$
(1.26)

Upon integrating

$$\phi = c_p \ln \theta + \text{const.},$$ (1.27)

which associates entropy with potential temperature. It is evident from the defining equation (1.25) that adiabatic processes ($dq = 0$) are also isentropic processes.

Meteorological thermodynamic charts

(a) Stüve diagram

The Stüve diagram (or simply "adiabatic chart") is a thermodynamic diagram based on the adiabatic equation (1.24). This equation shows that, for a given value of θ, there is a linear relation between T and p^k. Consequently, adiabatic processes follow straight line paths on a thermodynamic diagram with coordinates of T versus p^k.

This kind of chart is convenient for depicting adiabatic processes in the atmosphere. A line along which $\theta = \text{const.}$ is called an *adiabat*. Figure 1.4 is a schematic diagram of a Stüve chart, illustrating the working coordinates of pressure and temperature, and also the appearance of isobars, adiabats, and isotherms.

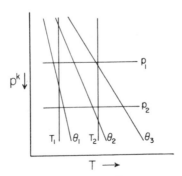

FIG. 1.4. Stüve diagram.

(b) Emagram

So-called *true* thermodynamic diagrams are ones on which area is proportional to energy (or work) with the same constant of proportionality over the whole chart. Thus a p,α-diagram is a true thermodynamic diagram, because the area in any closed contour is proportional to the work done in a cyclic process defined by the contour.

In meteorology the state variables most frequently employed to describe the air are pressure and temperature. It is possible to construct a true thermodynamic diagram with coordinates of p and T on the basis of (1.9) and (1.15). We have

$$dw = pd\alpha = R'dT - \alpha dp$$

and, for a cyclic process,

$$\int_c dw = \int_c R'\,dT - \int_c R'T\frac{dp}{p}. \tag{1.28}$$

But in (1.28) $R'dT$ is an exact differential which integrates to zero, so that the work done reduces to

$$\int_c dw = -R'\int_c Td(\ln p). \tag{1.29}$$

This result indicates that a chart with coordinates of T versus $\ln p$

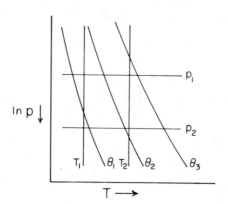

FIG. 1.5. Emagram.

has the properties of a true thermodynamic diagram. Such a chart is called an emagram, an abbreviation for energy-per-unit-mass diagram, and is illustrated schematically in Fig. 1.5.

(c) *Tephigram*

From the defining equation of entropy, it follows that the total heat added in a cyclic process is

$$\int_c dq = \int_c T d\phi = c_p \int_c T d(\ln \theta). \tag{1.30}$$

Consequently a chart with coordinates of T versus ϕ, or equivalently T versus $\ln \theta$, has the required area–energy relation of a true thermodynamic diagram. Such a chart is called a tephigram, standing for T,ϕ-diagram, and is shown schematically in Fig. 1.6.

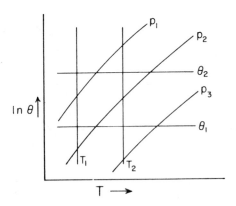

FIG. 1.6. Schematic tephigram.

The tephigram is usually rotated so that the isobars end up more or less horizontal with pressure decreasing upwards on the chart. Figure 1.7 illustrates a chart with this orientation. It is based on the Canadian meteorological service tephigram and will be used in the examples that follow.

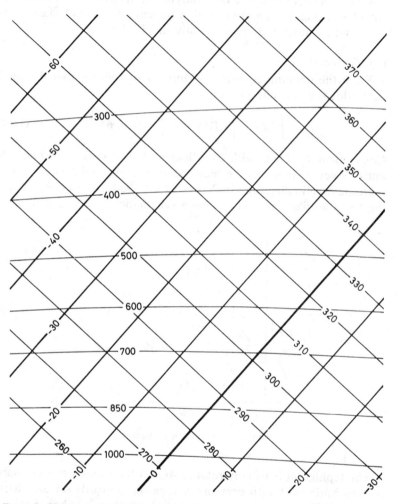

Fig. 1.7. Skeleton of a tephigram. Isobars are approximately horizontal, labeled in mb. Isotherms in deg C go upwards and to the right. Dry adiabats are normal to the isotherms and are labeled according to potential temperature (deg K).

Problems

1. The following table gives the approximate percentages by mass of the atmosphere's main permanent gases. Using these data, show that the effective molecular weight of dry air is 28.96 g/mole.

Gas	Mol. wt.	Mass, %
Nitrogen	28.016	75.57
Oxygen	32.000	23.15
Argon	39.944	1.28

2. A unit mass of dry air expands from an initial pressure of 500 mb to a final pressure of 400 mb.

(a) Calculate the work done assuming an isothermal expansion at $T = 300°K$.
(b) Calculate the amount of heat added in this process.
(c) Calculate the work done assuming an adiabatic expansion at $\theta = 300°K$.
(d) Calculate the change in temperature in this process.

3. A 50-g sample of air is initially at a pressure of 1000 mb and a temperature of 180°K. Heat is added isobarically until the sample expands by 10% of its original volume. Compute the final temperature of the sample, the work done in expanding, and the amount of heat added.

4. Consider a cyclic process in which a unit mass of dry air, initially at 0°C, goes through the following steps:

(1) Isothermal expansion from 700 to 600 mb.
(2) Isobaric cooling to −10°C.
(3) Isothermal compression back to 700 mb.
(4) Isobaric heating to 0°C.

Calculate the work done in this process. Confirm your answer using a tephigram or a similar thermodynamic chart.

5. The fundamental coordinates of a tephigram are T and $\ln \theta$. These axes are scaled in such a way that isobars become approximately horizontal when the whole chart is tilted about 45° in a clockwise sense. The isobars are not straight lines, however, and consequently they are not horizontal except at the particular point where their slope equals zero. Prove that the slope of the isobar through any point on the tephigram depends only on temperature.

6. The coordinates x and y of any thermodynamic chart are of the form $x = x(p,\alpha)$ and $y = y(p,\alpha)$. For example, in the case of an emagram $x = p\alpha/R' = T$ and $y = -R' \ln p$. A true thermodynamic diagram is one whose coordinates (x,y) may be transformed to (p,α) with a Jacobian of unity. That is, a sufficient condition that a given thermodynamic chart be a true thermodynamic diagram is

$$J\left(\frac{x,y}{p,\alpha}\right) = 1.$$

(a) Prove that a Stüve diagram is *not* a true thermodynamic diagram.
(b) How would you construct a true thermodynamic diagram with working coordinates of temperature and density? Indicate such a diagram schematically and sketch the approximate appearance of isobars and dry adiabats.

CHAPTER 2

WATER VAPOR AND ITS THERMODYNAMIC EFFECTS

Equation of state for water vapor

Unlike other atmospheric constituents, water appears in all three phases, solid, liquid, and vapor. In the vapor phase water in the atmosphere behaves as an ideal gas to a good approximation. Its equation of state is

$$e = \rho_v R_v T, \tag{2.1}$$

where e = vapor pressure, ρ_v = vapor density, and R_v = individual gas constant for water vapor (461 J $kg^{-1} K^{-1}$). This equation sometimes appears in the form

$$e = \rho_v \frac{R'}{\varepsilon} T, \tag{2.2}$$

where $\varepsilon = R'/R_v = m_v/m = 0.622$.

Water vapor has a specific heat at constant pressure equal to 1850 J $kg^{-1} K^{-1}$. The specific heat at constant volume equals 1390 J $kg^{-1} K^{-1}$.

Clausius–Clapeyron equation

Consider a closed and thermally insulated container partly filled with water as shown in Fig. 2.1. Molecules from the surface layer of water are in agitation and some break away as vapor molecules. On the other hand some of the vapor molecules collide with the surface and stick. Thus condensation and evaporation take place simultaneously. For a given temperature an equilibrium condition will eventually be reached when the two processes have the same rate. Then the temperature of the air and vapor equals that of the liquid and there is

15

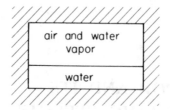

FIG. 2.1. Vapor in equilibrium with liquid surface.

no net transfer of molecules from one phase to the other. The space above the liquid is then said to be *saturated* with water vapor. The partial pressure due to the water vapor in this condition is called the saturation vapor pressure. It is found that saturation vapor pressure depends only on temperature, and this functional dependence is described by an important differential equation derived below.

Heat is required to change phase from liquid to vapor because the kinetic energy of the vapor molecules exceeds that of liquid molecules at the same temperature. We denote by L the heat required to convert a unit mass of liquid to vapor, with pressure and temperature held constant. This is the latent heat of vaporization. For this transition, from phase 1 (liquid) to phase 2 (vapor),

$$L = \int_{q_1}^{q_2} dq = \int_{u_1}^{u_2} du + \int_{\alpha_1}^{\alpha_2} p\,d\alpha = u_2 - u_1 + e_s(\alpha_2 - \alpha_1), \quad (2.3)$$

where e_s denotes the saturation vapor pressure, which is constant throughout the process. Since temperature is also constant, we may write as well

$$L = T \int_{q_1}^{q_2} \frac{dq}{T} = T(\phi_2 - \phi_1). \quad (2.4)$$

Equating results,

$$u_1 + e_s\alpha_1 - T\phi_1 = u_2 + e_s\alpha_2 - T\phi_2, \quad (2.5)$$

which shows that this particular combination of thermodynamic variables remains constant in an isothermal, isobaric change of phase. This combination is called the Gibbs function of the system and is denoted by G. Thus, for phase 1,

$$G_1 = u_1 + e_s\alpha_1 - T\phi_1 \qquad (2.6)$$

and (2.5) may be written simply as $G_1 = G_2$.

Though it is constant in the phase transition, the Gibbs function does vary with temperature and pressure, and its dependence on these variables may be determined by differentiation:

$$dG = du + e_s d\alpha + \alpha de_s - Td\phi - \phi dT. \qquad (2.7)$$

But $du + e_s d\alpha = dq = Td\phi$, and (2.7) reduces to

$$dG = \alpha de_s - \phi dT. \qquad (2.8)$$

Since G is the same for both phases, $dG_1 = dG_2$ and (2.8) implies

$$\frac{de_s}{dT} = \frac{\phi_2 - \phi_1}{\alpha_2 - \alpha_1} = \frac{L}{T(\alpha_2 - \alpha_1)}. \qquad (2.9)$$

This result expresses the dependence of saturation vapor pressure on temperature and is known as the Clausius–Clapeyron equation. Under ordinary atmospheric conditions $\alpha_2 \gg \alpha_1$ and water vapor behaves as an ideal gas. Then (2.9) reduces to

$$\frac{de_s}{dT} = \frac{L}{T\alpha_2} = \frac{Le_s}{R_v T^2}. \qquad (2.10)$$

If the latent heat were a constant, this equation could be integrated to give

$$\ln \frac{e_s}{e_{s0}} = \frac{L}{R_v} \left(\frac{1}{T_0} - \frac{1}{T} \right), \qquad (2.11)$$

where e_{s0} is the value of saturation vapor pressure at T_0, a constant of integration that must be determined by experiment or additional theoretical considerations. In fact, however, L depends weakly on temperature so that (2.11) is not an exact description of the dependence of e_s on T. Over the range of temperature encountered in the troposphere, L is always within a few per cent of 600 cal/g = 2.5×10^3 joules/g, and employing this value in (2.11) gives a useful approximation to the actual saturation vapor pressure. The constant of integration is $e_{s0} = 6.11$ mb at $T_0 = 273°K$.

At temperatures below 0°C, the equilibrium vapor pressure relative to ice is given by (2.10) with L replaced by L_s, the latent heat of sublimation.

Values of e_s and L over a range of temperature are given in Table 2.1. At temperatures of 0°C and colder the latent heat of sublimation and the equilibrium vapor pressure e_i over ice are also tabulated.

TABLE 2.1. *Latent Heat of Condensation (Sublimation) and Saturation Vapor Pressure for Various Temperatures. (Source, Smithsonian Meteorological Tables)*

T (°C)	e_s (mb)	e_i (mb)	L (cal g^{-1})	L_s (cal g^{-1})
−40	0.189	0.128	621.7	678.0
−35	0.314	0.223		
−30	0.509	0.380	615.0	678.0
−25	0.807	0.632		
−20	1.254	1.032	608.9	677.9
−15	1.912	1.652		
−10	2.863	2.600	603.0	677.5
−5	4.215	4.015		
0	6.108	6.107	597.3	677.0
5	8.719			
10	12.272		591.7	
15	17.044			
20	23.373		586.0	
25	31.671			
30	42.430		580.4	
35	56.236			
40	73.777		574.7	

Moist air: its vapor content

Atmospheric air is a mixture of dry air and water vapor. There are various ways of describing the vapor content.

(a) Vapor pressure e, the partial pressure of the water vapor.

(b) Vapor density ρ_v, also called absolute humidity, defined by (2.1).

(c) Mixing ratio w, defined as the mass of water vapor per unit mass of dry air.

$$w = M_v / M_d = \rho_v / \rho_d.$$

From the equation of state, $\rho_v = e/R_v T$ and $\rho_d = (p - e)/R'T$ so that

$$w = \varepsilon \frac{e}{p - e} \approx \varepsilon \frac{e}{p}. \tag{2.12}$$

(d) Specific humidity q, the mass of water vapor per unit mass of moist air.

$$q = \rho_v/\rho = \frac{\rho_v}{\rho_d + \rho_v} = \varepsilon \frac{e}{p - (1 - \varepsilon)e} \approx \varepsilon \frac{e}{p}. \qquad (2.13)$$

The saturation mixing ratio and saturation specific humidity, denoted by w_s and q_s, are defined by (2.12) and (2.13) by formally replacing e by e_s. Since $e_s = e_s(T)$, then w_s and q_s are functions of temperature and pressure only, and do not depend on the vapor content of the air. All meteorological thermodynamic charts contain "vapor lines", which are usually isopleths of w_s.

(e) Relative humidity f, the ratio of the mixing ratio to its saturation value, expressed in per cent.

$$f = 100 \frac{w}{w_s} \approx 100 \frac{e}{e_s}. \qquad (2.14)$$

(f) Virtual temperature T_v, the temperature of dry air having the same density as a sample of moist air at the same pressure.

For a sample of air of volume V, having total pressure p and vapor pressure e,

$$p = p_d + e = \rho_d \frac{R^*}{m_d} T + \rho_v \frac{R^*}{m_v} T$$

$$= \frac{R^* T}{V} \left[\frac{M_d}{m_d} + \frac{M_v}{m_v} \right]$$

$$= \rho R^* T \left[\frac{M_d}{m_d} + \frac{M_v}{m_v} \right] \frac{1}{M_d + M_v}$$

$$= \rho R' T \left[\frac{1 + w/\varepsilon}{1 + w} \right].$$

This result indicates that the equation of state for dry air may be applied to moist air if we include the correction factor in brackets. The virtual temperature is introduced to include this correction factor.

$$T_v = T \left[\frac{1 + w/\varepsilon}{1 + w} \right] \approx T[1 + 0.6w]. \qquad (2.15)$$

Thermodynamics of unsaturated moist air

(a) Gas constant

The equation of state for dry air can be applied to moist air if we formally replace T by T_v. Thus

$$p\alpha = R'T_v \qquad (2.16)$$

is a general equation of state applicable to dry or moist air. Frequently the difference between actual and virtual temperature may be neglected.

Alternatively the equation of state for moist air may be written

$$p\alpha = R_m T, \qquad (2.17)$$

where R_m is the individual gas constant for moist air, which must depend on the mixing ratio according to

$$R_m = R'[1 + 0.6w]. \qquad (2.18)$$

Obviously (2.16) and (2.17) are equivalent; it only depends on where you choose to apply the correction factor.

(b) Specific heat

To determine c_{vm}, the specific heat at constant volume for moist air, consider the addition of heat to a sample of air consisting of one gram of dry air plus w grams of water vapor.

$$(1 + w)dq = c_v dT + wc_{vv}dT,$$

where c_v is the specific heat of dry air and c_{vv} is the specific heat of the vapor. This shows that

$$c_{vm} = \left(\frac{dq}{dT}\right) = c_v \left[\frac{1 + wr}{1 + w}\right],$$

where

$$r = c_{vv}/c_v = 1.35/0.717 \approx 1.9.$$

Thus

$$c_{vm} \approx c_v[1 + 0.9w]. \qquad (2.19)$$

The same procedure may be employed to show that the specific heat at constant pressure for moist air is

$$c_{pm} \approx c_p [1 + 0.8w].$$ (2.20)

Combining (2.18) and (2.20), the exponent in the Poisson equation (1.23) for adiabatic processes in moist air is found to be

$$\frac{R_m}{c_{pm}} \approx k[1 - 0.2w].$$ (2.21)

Because w is of the order 10^{-2} or less, the correction factors in (2.18)–(2.21) may often be neglected.

Ways of reaching saturation

A sample of moist air may undergo a variety of processes that lead to saturation. Several of these processes are of theoretical importance, and introduce certain new temperatures that reflect the moisture content of the air.

(a) Dew point temperature T_d, defined as the temperature to which moist air must be cooled, with p and w held constant, in order that it just reach saturation with respect to water. (The frost point temperature is defined similarly, except for saturation relative to ice.) Clearly the saturation mixing ratio at the dew point equals the mixing ratio of the moist air: $w = w_s(T_d)$.

(b) Wet-bulb temperature T_w, defined as the temperature to which air may be cooled by evaporating water into it at constant pressure, until saturation is reached. (Note that w is not held constant, so that $T_d \neq T_w$ in general.)

Consider a sample of moist air consisting of one gram of dry air plus w grams of water vapor. The first law of thermodynamics for this sample in an isobaric process is (from 2.20)

$$dq = c_p dT[1 + 0.8w].$$

Associated with the evaporation of dw grams of water is a heat loss given by

$$(1 + w)dq = -Ldw.$$

Consequently

$$c_p dT = -Ldw \left(\frac{1}{1 + w}\right)\left(\frac{1}{1 + 0.8w}\right) \approx -Ldw[1 - 1.8w].$$

The correction factor may usually be neglected and

$$c_p dT = -L dw, \qquad (2.22)$$

which result describes the wet-bulb process. Neglecting the weak dependence of L on temperature, this equation may be integrated to yield

$$\frac{T - T_w}{w_s - w} = \frac{L}{c_p}. \qquad (2.23)$$

The wet-bulb temperature is thus expressible as a function of temperature and mixing ratio. w_s is the saturation mixing ratio at temperature T_w.

(c) Equivalent temperature T_e, defined as the temperature a sample of moist air would attain if all the moisture were condensed out at constant pressure. An expression for T_e follows from (2.23) if we set $w_s = 0$ (the final mixing ratio) and $T_w = T_e$. Thus

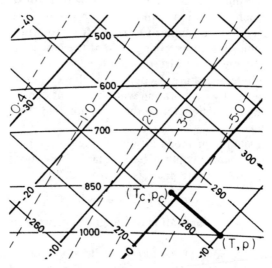

FIG. 2.2. Isentropic condensation pressure and temperature, indicated on a tephigram. Dashed lines are saturation mixing ratio, labeled in grams/kilogram. In the example shown the sample of air at 10°C, 1000 mb, has a mixing ratio of 4 g/kg with an isentropic condensation temperature of −1°C and an isentropic condensation pressure of 870 mb.

$$T_e = T + \frac{Lw}{c_p}.$$ (2.24)

(d) Isentropic condensation temperature T_c, defined as the temperature at which saturation is reached when moist air is cooled adiabatically with w held constant. This temperature is most readily understood with the help of a thermodynamic chart (Fig. 2.2).

The air initially has coordinates (T,p) with mixing ratio w. It is cooled adiabatically until its adiabat intersects the vapor line defined by $w_s = w$. The pressure at this intersection is called the isentropic condensation pressure, and the temperature is T_c.

Actually it is not obvious that condensation occurs once saturation is reached. Experience shows that this does happen in the atmosphere, so we can speak interchangeably of condensation point and saturation point.

Pseudoadiabatic process*

If expansion continues after the isentropic condensation point is reached, condensation occurs and the released latent heat will tend to warm the air. As a result, the temperature will decrease with pressure at a slower rate after condensation than before. To calculate the dependence of T on p in this process, assumptions must be made with regard to the condensed water. Does it stay with the air in the form of cloud droplets or does it precipitate out? At subfreezing temperatures is the condensate water or ice? The various alternatives are compared in standard texts on dynamic meteorology. It turns out that the final result—the dependence of T on p—is not significantly affected by the choice of assumptions. Consequently, only the simplest case is described here, the so called pseudoadiabatic process in which the condensate is assumed to be water which immediately precipitates. This is the simplest case

---

* In a brief history of early developments in the theory of the saturated adiabatic process, McDonald (1963a) explained that Lord Kelvin in 1862 gave the first correct description of the process; that Heinrich Hertz constructed an adiabatic diagram in 1884 that was the prototype of all subsequent meteorological thermodynamic charts; and that Wilhelm von Bezold in 1888 formulated the theory and equations for the pseudoadiabatic process.

because the heat content of the condensed material need not be considered when calculating temperature changes of the air. Also, the question of at what temperature sublimation becomes important is avoided.

Consider a sample of saturated air consisting of one gram of dry air plus w_s grams of water vapor. Let its pressure change by amount dp. The temperature will change by dT and a corresponding change dw_s in vapor content will occur. The equation relating these incremental changes is, to good approximation,

$$\frac{dT}{T} = k\frac{dp}{p} - \frac{L}{Tc_p}\,dw_s, \tag{2.25}$$

a mathematical description of the pseudoadiabatic process. This formula is the basis of "pseudoadiabats" on a thermodynamic chart.

Figure 2.3 illustrates the pseudoadiabatic expansion process. In an adiabatic expansion, the temperature decreases along a dry adiabat until the isentropic condensation point P is reached. Continued expansion is accompanied by the release of latent heat and the temperature follows along a pseudoadiabat from P onwards.

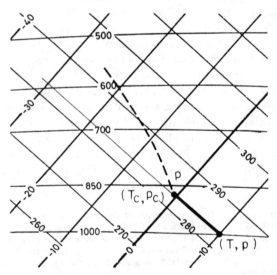

FIG. 2.3. Tephigram with pseudoadiabat.

Some additional special temperatures that may be defined by referring to this illustration are:

(a) Adiabatic wet-bulb temperature T_{sw}, obtained by following the pseudoadiabat from P down to the original pressure. The result is within 0.5°C of the wet-bulb temperature defined by (2.23).

(b) Wet-bulb potential temperature θ_w, defined by the intersection of the pseudoadiabat through P with the isobar $p = 1000$ mb.

(c) Equivalent temperature T_e (adiabatic definition), obtained by following up the pseudoadiabat from P to very low pressure, thus condensing out all the water vapor, and then returning to the original pressure along a dry adiabat. This temperature follows from integ-rating (2.25) from initial temperature T to final temperature T_e, and may be shown to be approximately

$$T_e = T \exp \left[\frac{L w_s}{c_p T_c} \right]. \tag{2.26}$$

Whether determined graphically, from (2.26), or (2.24), the value of T_e turns out to be about the same.

(d) Equivalent potential temperature θ_e, defined as the temperature a parcel of air would have if taken from its equivalent temperature to a pressure of 1000 mb in a dry adiabatic process.

Problems

1. It is readily seen from a p,α-diagram for water that condensation occurs when water vapor is compressed isothermally. This kind of process is of no importance in the atmosphere; in fact, condensation usually occurs as a result of expansion in processes which may be approximated as adiabatic. Show that over the meteorological range of temperatures adiabatic expansion of pure water vapor will lead to condensation.

2. Home humidifiers operate by evaporating water into the air in the house, and thereby raising its relative humidity. Consider a house having a volume of 200 m³, in which the air temperature is initially 21°C and the relative humidity is 10%. Compute the amount of water that must be evaporated to raise the relative humidity to 60%. Assume a constant pressure process at 1010 mb in which the heat required for evaporation is supplied by the air itself. A meteorological chart such as the tephigram is useful in solving this problem.

3. The temperature change associated with a pseudoadiabatic expansion process is given by

$$\frac{dT}{T} = k\frac{dp}{p} - \frac{L}{Tc_p}\,dw_s.$$

Determine whether temperature can ever *increase* during pseudoadiabatic expansion and, if so, under what conditions.

4. Suppose that air is exhaled from the lungs saturated at body temperature. Determine the ambient air temperature below which condensation of the breath should always occur. Is this result consistent with your experience or intuition? If not, what reasons can you give for the discrepancy?

5. A sample of moist air has a temperature of 280°K at a pressure of 900 mb, with a mixing ratio of 5 g/kg. Compute the following quantities for this sample:

(a) virtual temperature
(b) absolute humidity
(c) relative humidity
(d) dew point temperature

(e) wet-bulb temperature
(f) potential temperature
(g) equivalent temperature
(h) equivalent potential temperature

CHAPTER 3

STATIC STABILITY AND PARCEL BUOYANCY

Hydrostatic equilibrium

Air is said to be in hydrostatic equilibrium when it experiences no net force in the vertical direction. In this situation the vertical pressure gradient force on the air exactly balances the force of gravity (Fig. 3.1). For equilibrium

$$\frac{\partial p}{\partial z} = -\rho g, \tag{3.1}$$

the well-known hydrostatic equation.

Substituting for ρ from the equation of state gives

$$\frac{dp}{p} = -\frac{g}{R'T_v} \, dz. \tag{3.2}$$

In integrated form, we get

$$p = p_0 \exp\left[-\frac{g}{R'T_v}(z - z_0)\right], \tag{3.3}$$

where p is pressure at height z, and $\overline{T_v}$ is the mean virtual temperature over the pressure interval from p to p_0, given by

$$\overline{T_v} = \frac{\int_{\ln p_0}^{\ln p} T_v d(\ln p)}{\ln p - \ln p_0}. \tag{3.4}$$

An altitude scale can be constructed on a thermodynamic chart, using (3.3) to convert from pressure to height. Although the exact relation between z and p depends on temperature, approximate height scales are plotted on most charts based on a "standard

27

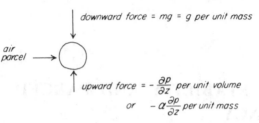

FIG. 3.1. A parcel of air in hydrostatic equilibrium.

atmosphere", which describes the average dependence of temperature on altitude.

Dry adiabatic lapse rate

For dry air undergoing an adiabatic change of pressure,

$$c_p dT = \frac{R'T}{p} dp.$$

Thus for dry air ascending and expanding

$$\frac{dT}{dz} = \frac{R'}{c_p} \frac{T}{p} \frac{dp}{dz}. \tag{3.5}$$

The pressure in an unconfined sample (parcel) of air will immediately adjust to the ambient pressure, so

$$\frac{dp}{dz} = \frac{\partial p}{\partial z} = -\rho' g, \tag{3.6}$$

where ρ' denotes ambient density:

$$\rho' = \frac{p}{R'T'} \tag{3.7}$$

with T' the ambient temperature.

Combining these equations shows that

$$\frac{dT}{dz} = -\frac{g}{c_p} \frac{T}{T'}.$$

Since the temperature of the parcel is not too different from ambient temperature, $T/T' \approx 1$ and the result simplifies to

$$\frac{dT}{dz} = -\frac{g}{c_p} \equiv -\Gamma, \qquad (3.8)$$

where $\Gamma = g/c_p = 0.98°C/100$ m $\approx 10^{-2}$ K/m denotes the *dry adiabatic lapse rate*. This is the rate at which temperature falls off with height in the process of dry adiabatic ascent. It can be shown that the adiabatic lapse rate for moist (but unsaturated) air is equal to Γ to a close approximation.

Buoyant force on a parcel of air

Consider a parcel of dry air with volume V having temperature T and density ρ. It displaces an equal volume of ambient air having temperature T' and density ρ'. The downward force on the parcel is equal to $\rho g V$. The downward force on the air displaced is equal to $\rho' g V$. The upward force is the same for parcel and displaced air, $-V(\partial p/\partial z)$. Hence the net buoyant force (upward) is $Vg(\rho' - \rho)$. Therefore, the buoyant force per unit mass is

$$F_B = g\left(\frac{\rho' - \rho}{\rho}\right) = g\left(\frac{T - T'}{T'}\right). \qquad (3.9)$$

As anticipated, this force is positive when the parcel is warmer than ambient air, negative when the parcel is cooler than ambient. For the case of moist air, (3.9) may be generalized by merely substituting virtual temperature for temperature.

Stability criteria for dry air

One of the uses of the dry adiabatic lapse rate is in assessing the stability of atmospheric layers with respect to the vertical displacement of a parcel. If after a small vertical displacement the parcel is subject to a restoring force which tends to accelerate it toward its original position, the layer is said to be stable. If after displacement the parcel is subject to a force in the direction of the displacement, the layer is said to be unstable. Whether a layer is stable or not depends on the ambient lapse rate, i.e. the decrease of temperature with height in the layer.

Consider a parcel of air with the ambient temperature T initially. If it is lifted adiabatically a small distance Δz it cools by the amount $\Gamma \Delta z$ and its temperature is reduced to $T - \Gamma \Delta z$. Let us denote the ambient lapse rate by γ, i.e.

$$-\left(\frac{\partial T}{\partial z}\right) = \gamma,$$

which is not to be confused with a *process* lapse rate. At height Δz above the initial position of the parcel, the ambient temperature is $T - \gamma \Delta z$. The excess temperature of parcel over ambient air is therefore $\Delta z(\gamma - \Gamma)$. When this quantity is positive the parcel is warmer than its surroundings and, by (3.9), is accelerated upwards. Consequently the air is unstable whenever $\gamma - \Gamma > 0$. Conversely the parcel is subjected to a restoring force (downward) whenever $\gamma - \Gamma < 0$. For the special case $\gamma = \Gamma$, the displaced parcel experiences zero buoyancy force. The stability criteria for dry air may thus be summarized

$$\gamma < \Gamma \qquad \text{STABLE}$$

$$\gamma = \Gamma \qquad \text{NEUTRAL}$$

$$\gamma > \Gamma \qquad \text{UNSTABLE}$$

These criteria may alternatively be expressed in terms of potential temperature. From the defining equation (1.24),

$$\frac{1}{\theta}\frac{\partial \theta}{\partial z} = \frac{1}{T}\frac{\partial T}{\partial z} - \frac{k}{p}\frac{\partial p}{\partial z} = \frac{1}{T}(\Gamma - \gamma), \qquad (3.10)$$

where we have employed (3.1), (3.8), and the equation of state. Equivalently, therefore, the stability conditions may be described by

$$\frac{\partial \theta}{\partial z} > 0 \qquad \text{STABLE}$$

$$\frac{\partial \theta}{\partial z} = 0 \qquad \text{NEUTRAL}$$

$$\frac{\partial \theta}{\partial z} < 0 \qquad \text{UNSTABLE}$$

The pseudoadiabatic lapse rate

Differentiating (2.25) with respect to height gives for the pseudoadiabatic process

$$\frac{dT}{dz} = \frac{kT}{p}\frac{dp}{dz} - \frac{L}{c_p}\frac{dw_s}{dz}. \tag{3.11}$$

By employing the hydrostatic equation and the Clausius–Clapeyron equation, this can be reduced to an expression for the pseudoadiabatic (or saturated adiabatic) lapse rate:

$$\Gamma_s \equiv -\frac{dT}{dz} = \Gamma \frac{\left[1 + \dfrac{Lw_s}{R'T}\right]}{\left[1 + \dfrac{L^2\varepsilon w_s}{R'c_pT^2}\right]}. \tag{3.12}$$

It can be seen from (3.12) that $\Gamma_s < \Gamma$ whenever $L\varepsilon > c_p T$. Owing to the high value of L for water, this inequality is always satisfied in the atmosphere.

Stability criteria for moist air

When a saturated parcel is displaced upwards its temperature will decrease at the pseudoadiabatic rate. If the ambient lapse rate is greater than pseudoadiabatic, the displaced parcel will find itself warmer than its surroundings and will be accelerated in the direction of the displacement. Such air is unstable with respect to pseudoadiabatic parcel displacement. Allowing for the possibility of condensation on ascent, five possible states of stability exist for moist air:

$$\gamma < \Gamma_s \qquad \text{ABSOLUTELY STABLE}$$
$$\gamma = \Gamma_s \qquad \text{SATURATED NEUTRAL}$$
$$\Gamma_s < \gamma < \Gamma \qquad \text{CONDITIONALLY UNSTABLE}$$
$$\gamma = \Gamma \qquad \text{DRY NEUTRAL}$$
$$\gamma > \Gamma \qquad \text{ABSOLUTELY UNSTABLE}$$

Convective instability

The weight of a column of air with unit cross sectional area which extends from pressure level p_1 up to p_2 is equal to $(p_1 - p_2)$. We consider vertical displacements of this column in which its weight remains constant. (Since g is constant to good approximation, this is equivalent to vertical displacements with mass remaining constant.) Under this condition, $\Delta p = p_1 - p_2$ is constant.

From the hydrostatic equation, the height of the column and its pressure-thickness are related by $\Delta p = g\bar{\rho}\Delta z$ with $\bar{\rho}$ the mean density of air in the column. Since ρ decreases with height, it follows that the lifting process considered here results in changes in Δz: stretching accompanies lifting of the column and contraction accompanies lowering. Usually the stability of the air will be affected by this process.

Consider first the case of dry air. Before displacement the stability of the air is measured by $\partial\theta/\partial z$ according to the criteria given previously. Thus over a small height interval δz the potential temperature varies by an amount equal to $\delta\theta = (\partial\theta/\partial z)\delta z$. For this incremental layer, $\delta\theta$ remains constant in adiabatic displacements. However, when the layer is lifted subject to the constraint described δz will increase in consequence of the stretching and it follows that $(\partial\theta/\partial z)$ must decrease. On the other hand $(\partial\theta/\partial z)$ must increase if the column of air is lowered. An exception to this is the case of air having neutral stability; in this case $(\partial\theta/\partial z) = 0$ before and after displacement.

These results mean that lifting does not affect the stability of an initially neutral layer. An initially unstable layer becomes less unstable; an initially stable layer becomes less stable. In short, lifting makes the lapse rate tend toward the dry adiabatic. Lowering a layer on the other hand makes its lapse rate depart further from adiabatic.

This effect may be readily illustrated on a thermodynamic chart. Shown in Fig. 3.2 is a layer of 100 mb thickness before and after lifting for the three possible stability conditions. It is noteworthy that air initially stable remains so after being lifted.

Lifting a column of moist air until it is saturated throughout also affects the stability. An important difference between dry and moist

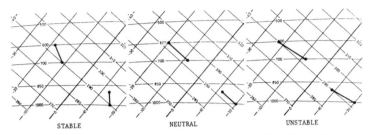

FIG. 3.2. Effect of lifting on stability of a layer (dry air case).

air is that moist air, initially stable, may be made absolutely unstable
or conditionally unstable by lifting. As shown above, this does not
happen for dry air.

A column of air which is rendered unstable by lifting to saturation
is said to be convectively unstable. (Some books use the term
potentially unstable.) The criteria for convective stability may be
expressed in terms of the lapse rate of wet-bulb potential temperature.

$$\frac{\partial \theta_w}{\partial z} > 0 \qquad \text{CONVECTIVELY STABLE}$$

$$\frac{\partial \theta_w}{\partial z} = 0 \qquad \text{CONVECTIVELY NEUTRAL}$$

$$\frac{\partial \theta_w}{\partial z} < 0 \qquad \text{CONVECTIVELY UNSTABLE}$$

These criteria are also best understood with reference to a tephigram
(Fig. 3.3).

Convective instability has to do with the lifting of layers and
should not be confused with conditional instability, which applies to
an undisplaced layer. A layer that is conditionally unstable need not
be convectively unstable; nor is a convectively unstable layer
necessarily conditionally unstable.

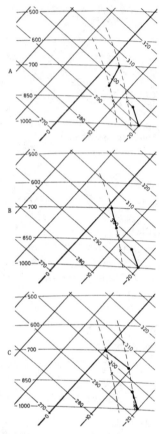

FIG. 3.3. Illustration of criterion for convective instability. In each case the layer between 900 and 1000 mb is lifted to 700 mb and condensation occurs throughout the layer. In case A, $\partial\theta_w/\partial z > 0$ and lifting to condensation stabilizes the layer. Case B has neutral convective stability with $\partial\theta_w/\partial z = 0$. Case C, $\partial\theta_w/\partial z < 0$, is convectively unstable.

Problems

1. The "homogeneous atmosphere" is a theoretical atmosphere throughout which the density is a constant. Unlike the real atmosphere, the homogeneous atmosphere has a well-defined "top" where the pressure falls to zero. Derive expressions for

   (a) the temperature lapse rate in a (dry) homogeneous atmosphere;
   (b) the height of the top of the homogeneous atmosphere.

2. A somewhat more general theoretical atmosphere is one with a constant lapse rate, $\partial T/\partial z = -B$. For this constant-lapse-rate atmosphere, derive expressions for

   (a) potential temperature as a function of pressure;
   (b) specific volume as a function of height.

3. The geopotential $\psi(h)$ is defined as the potential energy of a unit mass at height h above a reference level, usually mean sea level. By this definition $d\psi = gdz$ and $\psi(h) = \int_0^h gdz = gh$, where we have set $\psi(0) = 0$. Show that the geopotential at pressure level p is given by

$$\psi(p) = R'\overline{T_v}\ln(p_0/p),$$

where $\psi(p_0) = 0$.

4. In an unstable layer of air near the ground the temperature is found to decrease linearly with height at a rate of 2.5°C/100 m. A parcel of air at the bottom of this unstable layer with a temperature of 280°K is given an initial upward velocity of 1 m/sec. Assuming the parcel ascends dry adiabatically and encounters no resistance due to friction, show that after 1 min has elapsed it is approximately 80 m above the ground and ascending at 2.1 m/sec.

5. A sea-breeze is often established in coastal regions in the afternoon, as the air above the ground becomes warmer than that over the water. The cool air will tend to move landward at all altitudes where there is a horizontal pressure gradient towards the sea. The top of the sea-breeze may be considered as the altitude where this pressure gradient vanishes. Estimate the height of the sea-breeze in a case where the surface pressure is 2 mb greater over the water than over the land, and the temperature difference is 5°C. Take the surface pressure as 1000 mb and, for simplicity, assume isothermal temperature profiles over land and over water. Take the surface temperature over land as 15°C.

CHAPTER 4

MIXING AND CONVECTION

Mixing of air masses

(a) Isobaric mixing

Consider two masses of moist air at pressure p : the first with mass M_1, temperature T_1, and specific humidity q_1; the second with mass M_2, temperature T_2, and specific humidity q_2. Suppose these two samples are thoroughly mixed at constant pressure.

The specific humidity of the mixture is a mass-weighted mean of the individual specific humidities,

$$q = \frac{M_1}{M_1 + M_2} q_1 + \frac{M_2}{M_1 + M_2} q_2. \tag{4.1}$$

From (2.12) and (2.13) it follows that, to close approximation, the mixing ratio and the vapor pressure of the mixture are also weighted means,

$$w \approx \frac{M_1}{M_1 + M_2} w_1 + \frac{M_2}{M_1 + M_2} w_2 \tag{4.2}$$

$$e \approx \frac{M_1}{M_1 + M_2} e_1 + \frac{M_2}{M_1 + M_2} e_2. \tag{4.3}$$

Assuming that there is no net loss or gain of heat during mixing, the amount of heat lost by the warmer sample is equal to the amount of heat gained by the cooler sample. That is, if T denotes the temperature of the mixture,

$$M_1(c_p + w_1 c_{pv})(T_1 - T) = M_2(c_p + w_2 c_{pv})(T - T_2).$$

Neglecting the small contribution of water vapor to this balance,

$$T \approx \frac{M_1}{M_1 + M_2} T_1 + \frac{M_2}{M_1 + M_2} T_2, \tag{4.4}$$

36

showing that the temperature of the mixture is a weighted mean of the temperatures of the two samples.

This kind of mixing process is readily described with a hygrometric chart, which is a plot of e against T (see Fig. 4.1). Each of the two samples to be mixed is represented by a point on these coordinates, as illustrated. Equations (4.3) and (4.4) imply that the temperature and vapor pressure of the mixture will correspond to some position on the straight line connecting these points. This position will depend upon the ratio of M_1 to M_2. For example, if 4 kg of sample 1 are mixed with 3 kg of sample 2, the characteristic point of the mixture lies 4/7 the distance from point 2 up to point 1, as shown.

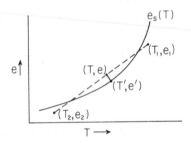

FIG. 4.1. Isobaric mixing of two air samples.

The solid line labeled e_s corresponds to the saturation vapor pressure, a function only of temperature according to the Clausius–Clapeyron equation (2.10). When two air masses are mixed under the conditions assumed here, the possibility exists that the mixture may be supersaturated, i.e. have a relative humidity above 100%. In this event condensation occurs and a cloud forms in the mixture.* The hygrometric chart may also be used to illustrate this process.

In the example shown, the mixture is supersaturated, so conden-

---

* From the observation that breath is visible in cold weather, and using an argument similar to that depicted in Fig. 4.1, the Scottish geologist James Hutton in 1784 deduced the concave-upwards shape of the curve which was described by the Clausius–Clapeyron equation a century later. This and many other interesting historical sidelights on cloud physics are described in a book by Middleton (1966).

sation will occur. Condensation will proceed until the mixture is just saturated. This will not occur at temperature T, because during condensation the mixture will be heated by the latent heat of condensation. As the mixing ratio decreases by amount dw during condensation, the latent heating is given approximately by

$$dq = -L\,dw. \tag{4.5}$$

Introducing (2.12) to convert from w to e, and assuming an isobaric process, the result is

$$\frac{de}{dT} = -\frac{pc_p}{L\varepsilon}, \tag{4.6}$$

which gives the slope of the line on a hygrometric chart describing the isobaric condensation process. The intersection of this line with the e_s curve defines a point with coordinates T', e', which corresponds to the mixture of air masses after condensation has occurred.

(b) Adiabatic mixing

Suppose the two samples of air considered above start with different pressures and are mixed after being brought adiabatically to the same pressure. As before, the specific humidity of the mixture is given by (4.1). In this adiabatic mixing process, the potential temperature of the mixture is a weighted mean of the potential temperatures of the two samples, in just the way that temperatures are related by (4.4).

Consequently, when a column of air is thoroughly mixed, the specific humidity will tend to a constant value throughout, given by

$$q_m = \frac{1}{M} \int_{z_1}^{z_2} \rho q\,dz,$$

where $M = \int_{z_1}^{z_2} \rho\,dz$ is the total mass of the column.

This equation may be written in terms of the pressure-thickness of the column by introducing the hydrostatic equation, leading to

$$q_m = \frac{1}{\Delta p} \int_{p_2}^{p_1} q\,dp. \tag{4.7}$$

Similar expressions apply to the mixing ratio and the vapor pressure of the mixture.

The potential temperature of the mixture tends to a constant value of

$$\theta_m = \frac{1}{\Delta p} \int_{p_2}^{p_1} \theta \, dp. \qquad (4.8)$$

With thorough mixing the temperature lapse rate in a vertical column thus approaches the dry adiabatic and the mixing ratio approaches a constant value; the limiting values of potential temperature and mixing ratio are averages with respect to pressure.

Convective condensation level

Vertical mixing within a column of air next to the ground occurs in consequence of solar heating of the surface. Heat is transferred by conduction from the surface to the air layer in contact with it, causing a strong lapse rate of temperature in the lowest layers of air. When the lapse rate becomes superadiabatic, any small disturbance will lead to vertical motions of elements of air in the layer, causing general mixing and overturning. The temperature profile in the mixing layer will tend toward the dry adiabatic and the mixing ratio will everywhere approach its average value with respect to pressure. If strong heating at the surface continues, this heat will be convected upwards, raising the potential temperature of the air throughout the layer. This process is indicated schematically in Fig. 4.2.

The heavy line indicates the initial temperature profile, with T_0 the

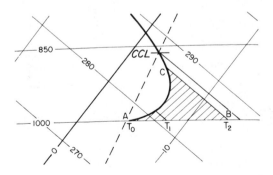

FIG. 4.2. Illustration of convective condensation level.

surface temperature. Heating raises this temperature causing a superadiabatic lapse rate temporarily and convection then tends to establish a dry adiabatic lapse rate with surface temperature T_1. Additional heating raises the temperature and increases the thickness of the mixing layer. Eventually the surface temperature reaches T_2 and the total heat added by convection is proportional to the hatched area ABC. Meanwhile the mixing ratio has become approximately constant up to the altitude of point C and equal to the average in the layer. The dashed curve indicates the profile of T_d, assuming constant w. It can be seen that a small amount of additional heating will raise the top of the mixing layer to the point where the adiabat intersects the vapor line. Condensation is expected at this point, which is called the convective condensation level (CCL). This is the level where you would expect to find the bases of cumulus clouds. If the air is conditionally unstable above the CCL, ascent of rising convective elements would continue with temperature falling off at the pseudoadiabatic lapse rate. This schematic picture illustrates the principle of cloud formation by convection, but the real process may often be complicated by the fact that heating causes evaporation of surface water which will increase the value of mixing ratio, tending to lower the CCl.

Related to the CCL is the lifting condensation level (LCL) which is defined as the level where a parcel of air just reaches saturation if it is lifted dry adiabatically from the surface. The LCL depends only on the properties of the air at the surface and does not provide for any vertical mixing. In conditions under which cumulus clouds are observed to form, the LCL and CCL often agree closely with one another because the air below cloud base is well mixed.

Adiabatic liquid water content

Consider a parcel of air at cloud base consisting of one gram of dry air and w_s grams of water vapor. (w_s is the saturation mixing ratio at the temperature and pressure of the cloud base.) As the parcel rises a small distance dz the saturation mixing ratio decreases by amount dw_s. If it is to remain just saturated, $-dw_s$ grams of water condense. Therefore

$$-\frac{dw_s}{1 + w_s} = d\chi$$

grams of water per gram of air condense. Associated with this condensation is the release of latent heat in the amount (per gram of air)

$$dq = -\frac{L}{1 + w_s} dw_s.$$

But $dq = c_p dT - \alpha dp$, so that $d\chi = (1/L)[c_p dT - \alpha dp]$. This result may be written

$$d\chi = \frac{1}{L}\left[c_p \left(\frac{dT}{dz}\right)_{sat.} dz - \alpha \frac{dp}{dz} dz \right],$$

which reduces to

$$d\chi = \frac{c_p}{L}(\Gamma - \Gamma_s)dz. \qquad (4.9)$$

The quantity χ here is called the adiabatic liquid water content of the air, in grams of water per unit mass of air. It represents the mass of cloud water under the assumptions that there is no mixing of the ascending parcel with ambient air and that none of the condensed water falls out as precipitation. It provides a useful estimate of the upper limit of cloud water content.

Often it is more convenient to measure adiabatic liquid water content in grams per unit volume of air instead of per unit mass. This quantity is denoted by M_a and related to χ by

$$M_a = \rho\chi.$$

The differential equation for M_a analogous to (4.9) is

$$\frac{dM_a}{dp} = \frac{M_a}{p}\left[1 - k\frac{\Gamma_s}{\Gamma} \right] - \frac{1}{L}\left[1 - \frac{\Gamma_s}{\Gamma} \right]. \qquad (4.10)$$

Although the dependence of Γ_s on temperature may in principle be obtained from (3.12), and (4.10) thereby solved for M_a, in practice one often uses a tephigram rather than equations to estimate adiabatic liquid water content.

Convection: elementary parcel theory

Convection refers to the vertical motions of elements of air. These motions can arise from buoyant or mechanical forces, and are the atmosphere's way of providing for efficient vertical transport of heat, mass, and momentum. Of special interest is buoyant convection, for this is the process leading to the formation of cumulus (or convective) clouds. Buoyant convection represents a conversion of potential energy to kinetic energy, and is expected to occur whenever heating at the surface or cooling aloft creates an unstable air layer.

Stability criteria were given in Chapter 3; an earlier section of the present chapter included a description of the way convection modifies the lapse rate in unstable air by transporting heat upwards. Not considered was the actual structure of the field of vertical air motion. Of importance are the sizes and shapes of the buoyant elements, their velocities, and their interaction with the surrounding air. These details of convection have been studied experimentally and theoretically by a large number of investigators and from various points of view over the past century.

Attention was first directed to the motion field established in incompressible fluids when heated uniformly at the bottom surface. This work was later extended by considering the effect of uniform or sheared motion of the fluid over the heated surface. More recent laboratory experiments with convection in fluids have shown striking similarities with cumulus clouds in the atmosphere and have led to theories of convection that account for some of the observed characteristics of clouds. Clouds themselves have been investigated by time lapse photography, instrumented airplanes, and radar to get an idea of the character of airflow in natural convection. Recently perfected techniques for remotely probing the atmosphere with acoustic or electromagnetic waves even permit the study of convective patterns in cloud-free air. Much of this work, though relevant to cloud physics, is beyond the scope of this book. However, some aspects of the physics of convection are presented in this and the following section, and later chapters treat the observation and numerical simulation of clouds. The interested reader is referred to Scorer's (1958) *Natural Aerodynamics* for a comprehensive view of the dynamics of convection and to Battan's (1973) *Radar Observation of the Atmosphere* for a summary of radar studies of clear-air convection.

The most elementary approach toward finding the vertical velocity of a convective element is based on (3.9), which expresses the buoyant force on a parcel of air. It is assumed that the parcel—a buoyant element of air with size and shape unspecified—maintains its identity in thermodynamic processes; that it in no way disturbs or interacts with the environmental air; that it has uniform properties throughout; that its pressure instantly adjusts to the pressure of the surroundings. Following from (3.9), its equation of motion is

$$\frac{d^2z}{dt^2} = gB, \qquad (4.11)$$

where

$$B = \frac{T - T'}{T'} \qquad (4.12)$$

represents the buoyancy term. Introducing $w = dz/dt$ as the vertical velocity, (4.11) becomes

$$wdw = gBdz, \qquad (4.13)$$

which may be integrated over height to give

$$w^2 = w_0^2 + 2g \int_{z_0}^{z} B(z)dz, \qquad (4.14)$$

where w is the velocity at height z and w_0 is the velocity at z_0.

The integral $\int_{z_0}^{z} gBdz$ may be shown using the hydrostatic equation to be equal to $R' \int (T - T')d(\ln p)$. It represents the area on a thermodynamic chart bounded by the process curve of parcel temperature and the ambient temperature profile, from pressure $p(z_0)$ up to pressure $p(z)$. This area is proportional to the increase in kinetic energy of the buoyant parcel between z_0 and z. It is referred to as the "positive area" of the sounding.

The velocity predicted by (4.14) is likely to be too high for the following reasons:

1. Aerodynamic drag was neglected.
2. Mixing with ambient air was neglected.

3. Compensating downward motions of the surrounding air were neglected.

4. The weight of condensed water, some of which is carried along with the parcel, was neglected.

Therefore w in (4.14) may be interpreted as an expected upper limit for vertical velocity in buoyant convection.

Modification of the elementary theory

In order to describe more accurately the behavior of convective elements or "thermals", the elementary parcel theory has been modified in various ways, in part as an attempt to overcome the shortcomings listed above. Some of these modifications are outlined in this section.

(a) The burden of condensed water

The buoyancy force per unit mass of air is gB for dry air, where B is given by (4.12). For moist air, the same expression holds if virtual temperature is used in place of temperature. If condensed water is present in the parcel, in the form of cloud droplets or precipitation, it exerts a downward force on the parcel equal to its weight. The buoyancy factor B then becomes

$$B = \frac{T}{T'} - (1 + m),\qquad(4.15)$$

where m denotes the "mixing ratio" of condensate, in terms of mass per unit mass of air. For the case of adiabatic expansion with no mixing, and neglecting precipitation, $m = \chi$, the adiabatic liquid water content given by (4.9). The expression (4.15) assumes that there is no condensate in the ambient air around the thermal. This would not be the case if the thermal were ascending through a cloud. A more general expression, allowing for this possibility, is

$$B = \frac{T}{T'}(1 + m') - (1 + m),\qquad(4.16)$$

where m' is the mixing ratio of condensate in the ambient air.

(b) Compensating downward motions

By the requirement of mass continuity, air must descend somewhere to replace the volume vacated by an upward-moving thermal. If the descending air is cloud-free, it will be warmed at the dry adiabatic rate. The air through which a thermal is ascending may thus have its temperature affected by adiabatic descent; the changes in temperature will then influence the buoyancy factor B.

The "slice-method" of stability analysis is designed to take into account this effect of ambient air descent. Attention is focused on a horizontal level through which thermals ascend and ambient air descends. The area occupied by thermals is denoted by A and the remainder, in which air is descending, is denoted by A'. The mass flux of upward-moving air through the level is $\rho w A$ where w is the velocity of the thermals. The downward mass flux is $\rho' A' w'$. It is assumed that the level is broad enough so that the upward and downward fluxes are equal. Therefore

$$\frac{A}{A'} = \frac{\rho' w'}{\rho w} \approx \frac{w'}{w}, \qquad (4.17)$$

where it is assumed that $\rho' \approx \rho$.

It is further assumed that the ascending air is cooled at the pseudoadiabatic rate while descending air is warmed at the dry adiabatic rate. Then, after a short time dt, the air arriving at the level from below will have a temperature given by $T_0 + \gamma w dt - \Gamma_s w dt$ where T_0 is the initial temperature at the level, Γ_s is the pseudoadiabatic lapse rate, and γ is the ambient lapse rate. The air arriving at the level from above will have temperature $T_0 - \gamma w' dt + \Gamma w' dt$. The situation is unstable when this temperature is less than the temperature of the thermal. Thus, for instability, we have

$$(\gamma - \Gamma_s)w > (\Gamma - \gamma)w'$$

or, from (4.17),

$$(\gamma - \Gamma_s)A' > (\Gamma - \gamma)A. \qquad (4.18)$$

In the limit as $A \to 0$, this result reduces to the instability criterion of an elementary parcel, as given earlier.

Using the slice-method arguments shows further that the neutral case arises whenever

$$\frac{\gamma - \Gamma_s}{\Gamma - \gamma} = \frac{A}{A'}.$$

If $A/A' > 0$ (that is, if the thermals are not negligible in size), this equation can only be satisfied if $\gamma > \Gamma_s$. Thus the lapse rate must be steeper for instability when the compensating downward motions are taken into account than when they are neglected.

(c) Dilution by mixing

When a buoyant element ascends, some mixing is expected to take place through its boundaries. Since the ambient air is generally cooler and drier than the buoyant element, the mixing will tend both to reduce the buoyancy of the thermal and to lower its mixing ratio. This kind of mixing is called "entrainment", and a theory has been developed to account for its thermodynamic effects.

Consider a mass of cloudy air m, consisting of dry air, water vapor, and condensed water. As it ascends through height dz, a mass dm of outside air is entrained. Let the cloudy air have temperature T, the ambient air T'. The heat required to warm the entrained air is then

$$dQ_1 = c_p (T - T')dm,$$

where the heat contents of the vapor and the condensate are assumed negligible compared to that of the dry air.

It is next assumed that just enough of the condensate evaporates to saturate the mixture. The heat required for this evaporation is

$$dQ_2 = L(w_s - w')dm,$$

where w' is the mixing ratio of the entrained air.

Condensation occurs during the ascent however, releasing an amount of heat given by

$$dQ_3 = -mLdw_s.$$

During the process, therefore, the cloudy parcel loses heat in the amount $dQ_1 + dQ_2$ and gains heat in the amount dQ_3. From the first law of thermodynamics,

$$m\left(c_p dT - R' T \frac{dp}{p}\right) = -(dQ_1 + dQ_2) + dQ_3.$$

By dividing both sides by $mc_p T$ and employing (1.26), it follows that in this entrainment process

$$\frac{d\theta}{\theta} = -\frac{L}{c_p T} dw_s - \left[B + \frac{L}{c_p T}(w_s - w')\right] \frac{dm}{m}. \qquad (4.19)$$

When there is no entrainment ($dm = 0$), this result reduces to an expression for the change of potential temperature in the pseudoadiabatic process, as expected. Since the bracketed term is always positive in cases of interest, (4.19) implies that temperature falls off at a faster rate when entrainment is taken into account than when it is neglected. This means that buoyancy is impaired by entrainment, as was anticipated.

(*d*) *Aerodynamic resistance: the bubble theory, jets and plumes*

Pictured in Fig. 4.3 is an idealized thermal, based primarily on laboratory studies of convection, although having a resemblance to atmospheric thermals which appear as protuberances or "turrets" on cumulus clouds.

Such thermals are observed to be shape-preserving, i.e. to have a form which maintains geometrical similarity during much of their development. A theory based on dimensional analysis has been

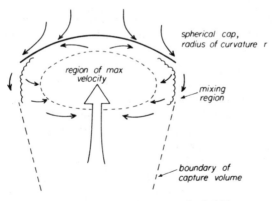

FIG. 4.3. Structure of a convective bubble.

formulated to explain some of their gross features. It is argued that the vertical velocity of a bubble depends on its size and the buoyancy according to the particular combination

$$w = c(g\bar{B}r)^{1/2}, \tag{4.20}$$

where w is the upward velocity of the cap, $\bar{B}$ is the average value of the buoyancy factor across the bubble, r is the radius of the cap, and c is a dimensionless constant to be determined experimentally or theoretically. Because of similarity the height of the cap above ground may be expressed as $z = nr$ and the volume of the thermal as $V = mr^3$, with n and m dimensionless constants to be determined. It is assumed that the total buoyancy (product of V times B) is conserved. Therefore, at any time,

$$V\bar{B} = V_0\bar{B}_0,$$

where V_0 and $\bar{B}_0$ denote the initial volume and buoyancy.

With these assumptions, (4.20) may be written

$$w = cn(gr_0^3\bar{B}_0)^{1/2}/z.$$

Upon integration this result reduces to

$$z^2 = 2cn\sqrt{\beta/m}\, t, \tag{4.21}$$

where $\beta = gV_0\bar{B}_0$. Laboratory results confirm (4.21), with $m \approx 3$, $n \approx 4$, and $c \approx 1.2$.

In the atmosphere, cumulus clouds are more complicated than the elementary bubbles, producible in the laboratory, which this theory is designed to explain. However, the individual spherical turrets in cumulus clouds are strongly suggestive of bubbles, and to a limited extent their behavior is consistent with the elementary theory. Their velocity is intimately connected with the stability of the air and the size and state of development of the cloud as a whole, and cannot be predicted for all clouds and for all occasions with the set of dimensionless parameters given above.

Even the elementary theory gives some insight on the interaction among thermals expected in cumulus clouds. Owing to the $\sqrt{t}$ law on height and size, successive thermals from the same place will tend to merge with one another, effectively increasing the buoyancy and rate of ascent of the (composite) thermal. If thermals do follow closely one after the other, the convection then takes the form of a continuous jet or plume instead of discrete bubbles. Again, laboratory experiments supported by a large body of theoretical work have been devoted to plumes, which are thought to approximate in some respects the airflow in developing cumulonimbus clouds.

An idealized plume is shown in Fig. 4.4. Its shape is conical, and the profiles of velocity and buoyancy across the plume are indicated. Since buoyancy and temperature are related through the factor B, the buoyancy profile is essentially the same as a temperature profile. Because of the conical shape, the radius is expressible as

$$r = \alpha z. \tag{4.22}$$

The mass flux through a given level is equal to $Awr^2\rho$, with units of g/sec, with A a dimensionless factor determined by the shape of the velocity profile. The momentum flux is $Aw^2r^2\rho$, with units of g cm sec^{-1}/sec, the same as force.

The flux of buoyancy is $c\rho gBwr^2$, with units of g cm sec^{-2}/sec, the same as force per unit time. c is a shape factor determined by the profiles of velocity and buoyancy. Of some theoretical interest is the pure buoyant plume, defined by

$$c\rho gBwr^2 = \text{const.}, \tag{4.23}$$

in which the buoyancy flux is constant over height. (The shape factor c is assumed constant.)

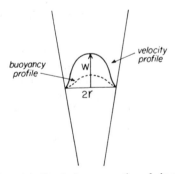

FIG. 4.4. Vertical cross-section of plume.

The buoyancy force and momentum are related in the following way: within a unit height interval the net buoyancy is $c\rho g B r^2$ and the momentum is $A w r^2 \rho$. The buoyancy in this slice of air is equal to the time rate of change of momentum. Thus

$$c\rho g B r^2 = \frac{d}{dt}(A w r^2 \rho) = w\frac{d}{dz}(A w r^2 \rho). \qquad (4.24)$$

Together, (4.23) and (4.24) may be employed to deduce the dependence upon height of the velocity and the buoyancy in the plume. It is assumed that these quantities are proportional to some power of height: $w \propto z^a$, $B \propto z^b$. Then the only way of satisfying (4.23) and (4.24) is for $a = -1/3$ and $b = -5/3$. Thus, for the case of a pure buoyant plume, theory predicts

$$w \propto z^{-1/3}; \qquad B \propto z^{-5/3}. \qquad (4.25)$$

The elementary theory presented here is for a dry plume, in which condensation is not considered. The closest atmospheric analogue to a plume is the updraft in cumulonimbus clouds, where condensation is occurring. The theory has been elaborated by several investigators to include the effects of condensation, and with these refinements it comes closer to describing natural convection. However, it still contains a number of adjustable dimensionless parameters that are not found to be the same under all conditions.

The most recent work on cloudy convection has not focused attention on the updraft only, nor employed arguments based on dimensional analysis, but has attempted to integrate the equations of motion of air while including the thermodynamics of condensation and precipitation production. Such dynamic models show the development of cloud and rain that closely resembles natural convection. Until quite recently, they have been either two-dimensional (at most) or three-dimensional only through assumptions of symmetry. Moreover, the nature of the assumed initial disturbance has been somewhat arbitrary. There are promising signs, however, that full three-dimensional, time varying models including precipitation will be perfected within the next few years. Some of the recent models of cloudy convection are described in Chapter 14.

Problems

1. Derive an expression for the change of relative humidity with respect to height. Prove that in a well-mixed, unsaturated layer of air the relative humidity will always increase with height. Assuming a well-mixed layer, evaluate the expression for $T = 250°K$ and $f = 50\%$.

2. In cloudy convection the mass of condensed water that is carried along with an ascending parcel of air exerts a downward force on it and thus reduces its buoyancy. This effect may be described by applying a correction factor to the parcel's temperature to account for the effect of condensed water on the buoyancy. (This is analogous to the virtual temperature correction to account for water vapor.) Derive an expression for the correction factor in terms of the mass of condensate present.

3. Suppose the ambient lapse rate is dry adiabatic, with a temperature of 280°K at 900 mb. Consider a parcel of saturated air at 900 mb with the ambient temperature. If this parcel is given an upward displacement it will be positively buoyant and will continue to ascend. Neglect entrainment and aerodynamic resistance and calculate the parcel's upward velocity at 700 mb assuming

   (a) elementary parcel theory without including the virtual temperature correction;
   (b) elementary parcel theory with allowance for virtual temperature;
   (c) parcel theory with a correction for the weight of condensed water, assuming full adiabatic water content.

4. Equal masses of two samples of air are thoroughly mixed and a fog is observed to form. The first sample has a temperature of 30°C with 90% relative humidity; the second a temperature of 2°C with 80% relative humidity. Assuming that the mixing occurs isobarically at 1000 mb, determine the temperature of the foggy air and its liquid water content, in grams of water per kilogram of air. Saturation vapor pressure as a function of temperature is given in the following table:

T (°C)	e_s (mb)
0	6.11
2	7.05
4	8.13
6	9.35
8	10.73
10	12.28
12	14.02
14	15.98
16	18.17
18	20.63
20	23.37
22	26.43
24	29.83
26	33.61
28	37.80
30	42.43

5. The amount of water vapor in the atmosphere is described by a quantity called precipitable water, which is defined as the mass of water vapor in a vertical column of the atmosphere having unit cross-sectional area. In the c.g.s. system, precipitable water has units of g/cm^2. Since 1 g of water occupies 1 cm^3, however, the units are usually expressed in cm^3/cm^2, or simply cm.

Derive an expression for the precipitable water between the surface and altitude H in terms of the mixing ratio. Assuming hydrostatic equilibrium, evaluate the precipitable water in a column between 1000 mb and 500 mb for a situation in which the mixing ratio varies linearly with pressure from 7 g/kg at 1000 mb to 1 g/kg at 500 mb.

6. The following sounding was obtained in central Alberta at a time when thunderstorms were developing.

Pressure (mb)	Temperature (°C)	Dew point (°C)
910 (surface)	23.5	14.5
850	17.0	12.5
800	12.5	10.8
770	10.0	6.0
745	10.0	−1.5
660	2.0	−10.0
555	−10.0	−13.0
525	−12.0	−19.0
500	−13.0	−18.0
400	−24.5	−30.5
300	−39.5	—
215	−55.0	—
190	−53.0	—
180	−49.0	—
125	−53.0	—

(a) Consider the air at the surface level. For this air, determine the following quantities:

(1) potential temperature   (6) equivalent temperature
(2) density   (7) equivalent potential temperature
(3) mixing ratio   (8) wet-bulb temperature
(4) relative humidity   (9) wet-bulb potential temperature
(5) virtual temperature   (10) isentropic condensation temperature

(b) Suppose a parcel of air from 910 mb ascends adiabatically.

(1) At what pressure would condensation be reached?
(2) Assuming ascent to continue beyond the condensation level, what would be the adiabatic liquid water content at 500 mb?

(3) What would be the increase in entropy of the parcel by the time 500 mb is reached?

(4) What would be the amount of latent heat released by the time 500 mb is reached?

(5) Estimate the vertical velocity of the parcel at 500 mb, neglecting friction and the mass of condensed water.

(c) What is the amount of precipitable water between the surface and 500 mb?

7. For a pure buoyant plume, it can be shown that the vertical velocity and the buoyancy factor of the ascending air vary with time according to $w \propto t^{\alpha}$ and $B \propto t^{\beta}$. Solve for the numerical values of α and β.

CHAPTER 5

FORMATION OF CLOUD DROPLETS

General aspects of cloud and precipitation formation

Phase changes of water play the central role in cloud microphysics. The possible changes are as follows:

vapor $\rightleftharpoons$ liquid   (condensation, evaporation)

liquid $\rightleftharpoons$ solid   (freezing, melting)

vapor $\rightleftharpoons$ solid   (deposition, sublimation)

Of primary meteorological importance are changes from left to right: these are changes in the direction of increased molecular order and are cloud forming processes. One of the intrinsic problems of cloud physics is that these phase transitions do not occur at thermodynamic equilibrium. These transitions, in the direction of increasing order, must overcome a strong "free energy barrier". Water droplets, for example, are characterized by strong surface tension forces. For a droplet to grow by condensation from the vapor, the surface tension has to be overcome by a strong gradient of vapor pressure.

The Clausius–Clapeyron equation describes the equilibrium condition for a thermodynamic system consisting of vapor and bulk water. Saturation is defined as the equilibrium situation in which the rates of evaporation and condensation are equal. Because of the free energy barrier of small droplets, phase transition does not generally occur at the equilibrium saturation of bulk water. In other words, if a sample of moist air is cooled adiabatically to the equilibrium saturation point, droplets should not be expected to form. In fact, water droplets begin to condense in pure water vapor only when the relative humidity (defined in terms of equilibrium conditions with respect to bulk water) reaches several hundred percent!

The classical problem of cloud physics is to explain why cloud droplets are observed to form in the atmosphere when ascending air just reaches equilibrium saturation. The answer to this question has been known for nearly a century, and rests in the fact that the atmosphere contains significant concentrations of particles of micron and submicron size which have an affinity for water and serve as centers for condensation. These particles are called condensation nuclei. The process by which water droplets form on nuclei from the vapor phase is called heterogeneous nucleation. The formation of droplets from the vapor in a pure environment, which requires high supersaturations and is not important in the atmosphere, is called homogeneous nucleation. All processes in which a free energy barrier must be overcome, such as vapor to liquid or liquid to ice transitions, are termed nucleation processes.

Many different types of condensation nuclei are present in the atmosphere. Some become wetted at relative humidities less than 100% and account for the haze that impedes visibility. The relatively large condensation nuclei are those which may grow to cloud droplet size. As moist air is cooled in adiabatic ascent, the relative humidity approaches 100%. The more hygroscopic nuclei then begin to serve as centers of condensation. If ascent continues, supersaturation is produced by the cooling and is depleted by the condensation on nuclei. By supersaturation is meant the excess of relative humidity over the equilibrium value of 100%. Thus air with a relative humidity of 101.5% has a supersaturation of 1.5%. In clouds, there are usually enough nuclei present to keep the supersaturation from rising much above about 1%. It is an important characteristic of the atmosphere that there are always enough condensation nuclei present to provide for cloud formation when the relative humidity barely exceeds 100%.

As a cloud continues to ascend, its top may eventually be cooled to temperatures below 0°C. The supercooled water droplets in the cloud may or may not freeze, depending upon whether ice nuclei are present. For pure water droplets, homogeneous freezing does not occur until a temperature of about −40°C is reached. When suitable nuclei are present, however, freezing can occur at just a few degrees below zero. Although these aerosols are not completely understood, it is significant that they are rather scarce in the atmosphere, unlike the abundant condensation nuclei. Consequently supersaturations of more than a few tenths of a percent are extremely uncommon in the atmosphere, although water droplets in supercooled form are the regular state of affairs. Supercoolings down to −15°C or colder are not uncommon. For this reason the principal means of artificially modifying clouds is the addition of ice nuclei, as we shall see later.

A cloud is an assembly of tiny droplets numbering in the order of 100 per cubic centimeter and having radii of about 10 μm. This structure is extremely stable as a rule and the droplets show little tendency to come together or to change their sizes except by general growth of the whole population. Precipitation is developed when the cloud population becomes unstable, and some drops grow at the expense of others. There are two mechanisms whereby a cloud microstructure may become unstable. The first involves the direct collision and coalescence (sticking) of water droplets and may be important in any cloud. The second mechanism involves the interaction between water droplets and ice crystals and is confined to those clouds whose tops penetrate above the 0°C level.

From analysis of the aerodynamic forces involved it is found that very small droplets cannot readily be made to collide. A small drop falling through a cloud of still smaller droplets will collide with only a minute fraction of the droplets in its path so long as its radius is less than about 18 μm. Therefore it is expected that clouds which contain negligible numbers of drops larger than 18 μm will be relatively stable with respect to growth by coalescence. Clouds with considerable numbers of larger drops may develop precipitation.

When an ice crystal exists in the presence of a large number of supercooled water droplets the situation is immediately unstable. The equilibrium vapor pressure over ice is less than that over water at the same temperature and consequently the ice crystal grows by diffusion of vapor and the drops evaporate to compensate for this. The vapor transfer depends on the difference in equilibrium vapor pressure of water and ice and is most efficient at about −15°C.

Once the ice crystal has grown appreciably larger than the water droplets, it begins to fall relative to them and collisions become possible. If the collisions are primarily with other ice crystals snowflakes form, and if water droplets are collected graupel or hail may form. Once the particle falls below the 0°C level melting can occur, and the particle may emerge from cloud base as a raindrop indistinguishable from one formed by coalescence. In cold weather, or when large hailstones are involved, the particle may of course reach the ground unmelted.

The particles of interest in cloud physics have a wide range of size, concentration, and fall velocity. Figure 5.1, from McDonald (1958), compares these properties for some of the particles involved in condensation and precipitation processes. Noteworthy are the vast differences in size between a typical condensation nucleus and a cloud drop, and between cloud drop and raindrop. To account for the production of natural rain the growth processes must be fast enough to get from condensation nucleus to raindrop in about 20 min.

Nucleation of liquid water in water vapor

The question of nucleation is: "How readily can an embryonic droplet be formed, by chance collisions and aggregations of water molecules, which is stable and which will continue to exist under the given environmental conditions?"

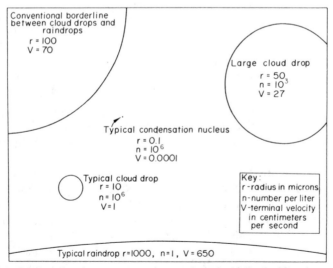

FIG. 5.1. Comparative sizes, concentrations, and terminal fall velocities of some of the particles involved in cloud and precipitation processes. (From McDonald, 1958.)

The droplet will be stable if its size exceeds a certain critical value. On the average, droplets larger than the critical size grow while those which are smaller decay. What determines the critical size is the balance between the opposing rates of growth and decay. These rates, in turn, depend upon whether the droplet forms in free space (homogeneous nucleation) or in contact with another body (heterogeneous nucleation). In the case of homogeneous nucleation of pure water, the growth rate depends upon the partial pressure of water vapor in the surroundings since this determines the rate at which water molecules impinge upon the droplet. The decay process, evaporation, depends strongly upon the temperature of the droplet and its surface tension since molecules at the surface of the drop must obtain a sufficient quantity of energy in order to overcome the binding forces if they are to escape. If equilibrium is established between the liquid and its vapor, the rates of condensation and evaporation are exactly balanced and the vapor pressure is equal to the equilibrium or saturation vapor pressure. But the equilibrium vapor pressure over a droplet's surface depends upon its curvature and is given by

$$e_s(r) = e_s(\infty) \exp(2\sigma/rR_v\rho_L T), \qquad (5.1)$$

where $e_s(r)$ is the saturation vapor pressure over the surface of a spherical droplet of radius r with surface tension σ and density ρ_L at temperature T. R_v is the gas constant for water vapor and $e_s(\infty)$ is the saturation vapor pressure over bulk water—the quantity most readily measured. Note that as the size of the drop decreases, the vapor pressure required for equilibrium with it increases. This equation was first derived by William Thomson (1870) (later Lord Kelvin), in slightly different form, to explain the rise of liquids in capillary tubes.

The surface tension σ is the free energy per unit surface area of the liquid. It can be defined as the work per unit area required to extend the surface of the liquid at constant temperature. Since work is the product of force × distance, work per unit area has units of force per unit length. The surface tension of water is about 75 dynes/cm over the meteorological range of temperature.

The net rate of growth of a droplet of radius r is proportional to the difference $e - e_s(r)$, where e denotes the actual ambient vapor pressure. Thus, drops with radii such that $e - e_s(r) < 0$ tend to decay while those with radii such that $e - e_s(r) > 0$ tend to grow. The critical size r_c is therefore the radius for which $e - e_s(r_c) = 0$ or $e = e_s(\infty) \exp(2\sigma/r_c R_v \rho_L T)$. Hence

$$r_c = \frac{2\sigma}{R_v \rho_L T \ln S}, \qquad (5.2)$$

where $S = e/e_s(\infty)$ is the saturation ratio. In order for a droplet, formed by chance collisions among water molecules, to be stable, it must grow to a radius exceeding r_c. The following table gives values of the critical droplet radius and the number of molecules in the critical droplet for various saturation ratios at 0°C.

TABLE 5.1. *Radii and Number of Molecules in Droplets of Pure Water in Equilibrium with the Vapor at 0°C*

Saturation ratio S	Critical radius r_c (μm)	Number of molecules n
1	∞	∞
1.01	1.208×10^{-1}	2.468×10^{8}
1.10	1.261×10^{-2}	2.807×10^{5}
1.5	2.964×10^{-3}	3.645×10^{3}
2	1.734×10^{-3}	730
3	1.094×10^{-3}	183
4	8.671×10^{-4}	91
5	7.468×10^{-4}	58
10	5.221×10^{-4}	20

It is seen that high supersaturations are required for very small droplets to be stable. When the supersaturation is 1%, corresponding to $S = 1.01$, droplets with radii smaller than 0.121 μm are unstable and will tend to evaporate.

In homogeneous nucleation, critical sized droplets are formed by random collisions of water molecules. If these capture another molecule, they become supercritical; that is, with increasing size, $e_s(r)$ decreases and the rate of growth (which is proportional to $e - e_s(r)$) increases. Supercritical droplets therefore grow spontaneously. The rate of nucleation is simply the rate at which supercritical droplets are formed and is given by the product of the concentration of critical droplets and the rate at which a critical droplet gains another molecule and becomes supercritical. By means of statistical thermodynamics (see for example Hirth and Pound, 1963, or Frenkel, 1946) the nucleation rate per unit volume is written as

$$J = 4\pi r_c^2 \frac{e}{\sqrt{2\pi mkT}} Zn \exp\left(-\frac{4\pi r_c^2 \sigma}{3kT}\right), \qquad (5.3)$$

where m is the mass of a water molecule, k is Boltzmann's constant, n is the number density of vapor molecules and Z is a factor which takes values in the order of 0.01. By substituting for r_c in terms of S, an expression results for the rate of homogeneous nucleation as a function of the saturation ratio at a given temperature. An important characteristic of this dependence is that the nucleation rate goes from undetectably small values to extremely large values over a very narrow range of S. The value of S for which this occurs is called the critical saturation ratio S_c and corresponds to $J = 1 \text{ cm}^{-3} \text{ sec}^{-1}$. This is readily measured. The experimental and theoretical values of S_c at 275.2°K are 4.2 ± 0.1 and 4.2 respectively and at 261.0°K are 5.0 ± 0.1 and 5.0 respectively. Such large saturation ratios are never observed in the atmosphere, supersaturations the order of less than a per cent being most common, and so homogeneous nucleation of liquid water from its vapor is of secondary importance to heterogeneous nucleation.

In the atmosphere cloud droplets form on the aerosols called condensation nuclei or hygroscopic nuclei. The rate of droplet formation is determined by the number of these nuclei present, and not by collision statistics. In general, aerosols can be classified according to their affinity for water as hygroscopic, neutral, or hydrophobic. Nucleation on a neutral aerosol requires about the same supersaturation as homogeneous nucleation. On a hydrophobic aerosol particle, which resists wetting, nucleation is more difficult and requires even higher supersaturations. But for the hygroscopic particles, which are soluble and have an affinity for water, the supersaturation required for droplet formation can be much less than its value for homogeneous nucleation.

A dissolved solute tends to lower the equilibrium vapor pressure of a liquid. Very roughly, the effect may be thought of as arising from the fact that, when solute is added to a liquid, some of the liquid molecules that were in the surface layer are replaced by solute molecules. If as is usually the case the vapor pressure of the solute is less than that of the solvent it follows that the vapor pressure is reduced in proportion to the amount of solute present. This effect can drastically lower the equilibrium vapor pressure over a droplet; the result is that a *solution droplet* can be in equilibrium with an environment at much lower supersaturation than a pure water droplet of the same size.

For a plane water surface the reduction in vapor pressure due to the presence of a solute may be expressed

$$\frac{e'}{e_s(\infty)} = \frac{n_0}{n + n_0},$$

where e' is the equilibrium vapor pressure over the solution, which consists of n_0 molecules of water and n molecules of solute. This is known as Raoult's Law. For dilute solutions

$$\frac{e'}{e_s(\infty)} = 1 - \frac{n}{n_0}. \tag{5.4}$$

For solutions in which the dissolved molecules are dissociated, (5.4) must be modified by multiplying n by the factor i, the degree of ionic dissociation. For NaCl, for example, $i \approx 2$ for dilute solutions. The number of effective ions in a solute of mass M is given by

$$n = iN_0M/m_s,$$

where N_0 is Avogadro's number (the number of molecules per mole) and m_s is the molecular weight of the solute. The number of water molecules in mass m may likewise be written

$$n_0 = N_0m/m_v.$$

Writing for the mass of water $m = \frac{4}{3}\pi r^3 \rho_L$, we note that (5.4) can now be expressed as

$$\frac{e'}{e_s(\infty)} = 1 - b/r^3, \tag{5.5}$$

where $b = 3im_vM/4\pi\rho_Lm_s$.

Kelvin's equation (5.1) and the solution effect are combined to give for the equilibrium saturation ratio e_s' (r) of a solution droplet

$$\frac{e_s'\,(r)}{e_s\,(\infty)} = \left[1 - \frac{b}{r^3}\right] e^{a/r},$$

where $a = 2\sigma/\rho_L R_v T$. For r not too small, a good approximation to this equation is

$$\frac{e_s'\,(r)}{e_s\,(\infty)} = 1 + a/r - b/r^3. \tag{5.6}$$

In this approximate form, a/r may be thought of as a "curvature term" which expresses the increase in saturation ratio over a droplet as compared to a plane surface. The term b/r^3 may be called the "solution term", as it shows the reduction in vapor pressure due to the presence of a solute. In c.g.s. units $a \approx 3.3 \times 10^{-5}/T$ and $b \approx 4.3 i M/m_s$. For given values of T, M, and m_s, (5.6) describes the dependence of saturation ratio on the size of a solution droplet. The resultant curve is called a Köhler curve, an example of which is illustrated in Fig. 5.2.

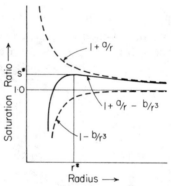

FIG. 5.2. Equilibrium saturation ratio as a function of size of solution droplet.

The curve shows that the solution effect dominates for small enough r, so that a very small solution droplet can be in equilibrium at relative humidities less than 100%. As the relative humidity is increased, the droplet will tend to grow until it is large enough for equilibrium again. This process of increasing the relative humidity and permitting the droplet to grow can be continued beyond the relative humidity of 100%. Finally the critical saturation ratio S^* is reached which corresponds to the peak of the Köhler curve. The corresponding equilibrium radius is r^*. Up to this point it was required that the relative humidity be increased in order for the droplet to grow. But notice now that once the droplet grows only slightly beyond r^* its equilibrium saturation ratio is less than S^*. Consequently the vapor will diffuse toward the droplet and it will continue to grow without additional increase in environment saturation ratio. Thus, droplets larger in size than r^* can grow without further increase in S. These droplets can grow to cloud droplet size. The droplets smaller than r^* grow only in response to increases in relative humidity and are termed "haze particles". A condensation nucleus is said to be "activated" when the droplet formed on it grows to size r^*. Once a droplet has exceeded this critical radius, growth will continue indefinitely according to the simple theory presented so far. In actual clouds this continued growth does not occur because there are many droplets present which compete for the available vapor and tend to lower the saturation ratio once the condensation becomes more rapid than the production of supersaturation by adiabatic ascent.

The critical values of radius and saturation ratio can be obtained from the approximate expression (5.6) and are given by

$$r^* = \sqrt{3b/a} \tag{5.7}$$
$$S^* = 1 + \sqrt{4a^3/27b}. \tag{5.8}$$

Table 5.2 gives examples of critical radii and supersaturations for droplets formed on sodium chloride nuclei.

TABLE 5.2. *Values of* r^* *and* $(S^* - 1)$ *as Functions of Nucleus Mass and Radius, assuming NaCl Spheres at a Temperature of 273°K*

Mass of dissolved salt (g)	r_s (μm)	r^* (μm)	$(S^* - 1)$ (%)
10^{-16}	0.0223	0.19	0.42
10^{-15}	0.0479	0.61	0.13
10^{-14}	0.103	1.9	0.042
10^{-13}	0.223	6.1	0.013
10^{-12}	0.479	19	0.0042

Atmospheric condensation nuclei

Natural aerosols, some of which are hygroscopic, range in size from about 10^{-3} μm radius for the small ions, made up of charged clusters of a few molecules, to more than 10 μm for the largest salt, dust, and combustion particles. Their concentrations vary over a wide range with respect to location and time. The small ions are not important in droplet formation, affording only a little improvement over homogeneous nucleation, while the 10 μm-size particles may not be too important because of their limited residence time in the atmosphere.

Particles up to 100 μm in size have been observed near the ground and even up to the altitude of cloud base during thunderstorm conditions (Rosinski, 1974). Stirred up by strong winds, these particles can remain only a short time before settling out. However, if ingested into cloud, they could play a role in precipitation development.

According to Brock (1972), about 75% of the total mass of aerosol material in the atmosphere is accounted for by natural and anthropogenic primary sources such as wind-generated dust (20%), sea spray (40%), forest fires (10%), and combustion and other industrial operations (5%). The remaining 25% is attributed to secondary sources which involve conversion of gaseous constituents of the atmosphere to small particles by photochemical and other chemical processes. Regardless of their mechanism of introduction, atmospheric aerosols continually undergo a variety of chemical and physical transformations, including coagulation, condensation, scavenging, washout, sedimentation, dispersion, and mixing. A comprehensive introduction to aerosol physics is given by Hidy and Brock (1970).

Aerosol size distributions, as those of cloud droplets, raindrops, or particulates in general, are usually described in terms of a distribution function $n(r)$ such that $n(r)dr$ equals the number of particles per unit volume of air whose radii lie in the interval between r and $r + dr$. Alternately, the cumulative distribution $N(r)$ may be used, defined by

$$N(r) = \int_r^\infty n(r')dr'$$

and expressing the number of particles per unit volume whose radii exceed r. Obviously

$$n(r) = -\frac{d}{dr} N(r).$$

Because of the wide range over which both n and r vary, a log-increment distribution function $n_1(r)$ is often employed for aerosols, defined by

$$n_1(r) = -\frac{d}{d(\log r)} N(r),$$

which equals the number of particles in the size interval $d(\log r)$.

Slinn (1975) has given a detailed, yet still idealized, presentation of typical aerosol spectra. Shown in Fig. 5.3, his distributions illustrate a variety of background and extreme conditions. He found that the most consistent way to compare such diverse data was by using log-area distributions $A(D)$, which express the total cross-sectional area of the aerosol particles in logarithmic increments of diameter. These distributions are related to the fundamental distribution $n(r)$ by

$$A(D) = \tfrac{1}{2}\pi D^3 n \left(\frac{D}{2}\right).$$

Thus, in a unit volume of air, $A(D)\, d\,(\ln D)$ is the total cross-section of aerosols whose diameters are in the interval $d \ln D$.

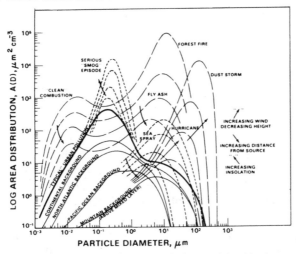

FIG. 5.3. Idealized aerosol spectra, showing typical ground-level background distributions and the general dependence of the spectra on height, wind speed, distance from source, and surface heating (From Slinn, 1975).

Although aerosol spectra are highly variable, it is useful to have approximate analytical descriptions or models characterizing their shape for such applications as diffusion studies, light scattering by the atmosphere, and infrared radiative transfer. One of the models frequently employed is the Junge distribution (after the atmospheric chemist who has reported a large number of aerosol measurements), defined by $n_1 (r) \propto r^{-3}$ or $dN/d (\log r) \propto r^{-3}$. This follows from the observation that the background aerosol spectrum, when plotted as a log-increment distribution function on logarithmic coordinates, can often be approximated by a straight line with a slope of -3 in the radius interval between approximately 10^{-1} and 10 μm.

Condensation nuclei, which make up some of the aerosol population, are broadly classified according to size as Aitken nuclei ($r < 0.2$ μm), large nuclei (0.2 μm $< r < 1$ μm), and giant nuclei ($r > 1$ μm). Typical concentrations of these nuclei in the lower troposphere are given in Table 5.3, condensed from data in Fletcher (1962). It is believed that the Aitken nuclei are primarily products of combustion and, to some extent, of natural reactions in the atmosphere. The large and giant nuclei are thought to be salt particles resulting from the bursting of surf

TABLE 5.3. *Concentrations of Nuclei (cm^{-3})*

	Aitken	Large	Giant
Over water	$10-10^2$	*	*
Over land	10^3-10^6	10^2	1

*Sea salt particles, with concentration depending on wind speed and surface roughness.

bubbles. Typical maritime clouds have droplet concentrations of a few tens per cm^3. Since this corresponds to the number of large salt nuclei observed with moderate winds, it seems that salt nuclei are in large part the responsible nuclei for maritime clouds. On the other hand clouds formed inland usually contain several hundred drops per cm^3, while the number of large nuclei identified as sea salt are only perhaps 10 per cm^3. It is believed that the responsible nuclei for land clouds might be large combustion produced nuclei and perhaps large nuclei which are coagulations of Aitken nuclei.

The continental aerosols are thought to have three main components in the size range above 0.1 μm. The first is sea salt, which is the predominant constituent of nuclei larger than 1 μm. The second is sulfate component, which may be present either as sulfuric acid or combined as a salt, perhaps ammonium sulfate, which predominates between 0.1 and 1 μm in diameter. The third component consists of insoluble particles probably derived from the soil, and its concentration depends on the condition of the soil and the average ground wind speed. The relative importance of these three components will depend upon the history of the airmass.

Condensation nuclei of some sort are always present in the atmosphere in ample numbers; clouds occur whenever there are vertical air motions and sufficient moisture. But as we shall see, in some marginal cases precipitation is more likely to occur when the nucleus population consists of a few large particles rather than many small particles.

Some of the measures of atmospheric nuclei are not relevant to cloud physics. For example, Aitken nucleus counts, as obtained with an expansion chamber counter, are essentially counts of *all* condensation nuclei in an atmospheric sample. In the expansion chamber supersaturations of several hundred percent are created so that essentially all nuclei present are activated. In the natural atmosphere, the only nuclei that enter into cloud forming processes are those that activate at supersaturations of about 1% and less. These in fact are now referred to as cloud condensation nuclei to distinguish them from condensation nuclei in general. What is important in cloud physics is the number of nuclei that become activated to form cloud droplets as a function of the supersaturation. Such "activity spectra" are measured by instruments developed during the past decade called diffusion–gradient cloud chambers. There are two types, the thermal gradient chamber and the chemical gradient chamber. They are capable of producing very slight supersaturations (values like several tenths of a percent) with a high degree of accuracy. An air sample is brought into the chamber having a known degree of supersaturation; by optical means the number of nuclei that grow to activation size are observed and counted. Then the supersaturation is increased a bit and the number of condensation centers is observed again. It has been found that the number of nuclei with critical supersaturation less than $(S - 1)$, per unit volume of air, can often be approximated by

$$N = C(S - 1)^k \qquad (5.9)$$

where C and k are parameters that depend upon the airmass. Jiusto (1966) has reported activity spectra for several localities which suggest that maritime conditions might be approximated by $C = 50$, $k = 0.4$, and continental conditions by $C = 4 \times 10^3$, $k = 0.9$. Measurements at any particular site might be expected to lie between these rather extreme cases. (The values quoted for C are appropriate for $(S - 1)$ expressed in per cent.)

Assuming an activity spectrum of the form (5.9), Twomey (1959) showed that the droplet concentration formed in an updraft of speed U can be expressed in terms of U, C, and k. For k between 0.4 and 1.0, his result may be approximated by

$$N \approx 0.88 C^{2/(k+2)} [7 \times 10^{-2} U^{3/2}]^{k/(k+2)} \qquad (5.10)$$

where N is in cm^{-3}, U is in cm/sec, and C is appropriate for percentage supersaturation. Twomey also obtained an expression for the peak supersaturation in the updraft, which may be approximated by

$$(S - 1)_{max} \approx 3.6[(1.6 \times 10^{-3} U^{3/2})/C]^{1/(k+2)}, \qquad (5.11)$$

with $(S - 1)$ expressed in per cent.

Comparisons of activity spectra with observed droplet spectra are beginning to provide experimental confirmation of the close relation between the nucleus population and the resulting cloud droplets. The development of a cloud after its formative stage, and in particular the amount and character of precipitation produced, is controlled more by large-scale phenomena, such as updraft speed and moisture supply, than by its microphysical structure. But to some extent the microstructure determines how susceptible a cloud is to producing precipitation, and how much time is required for the precipitation to form.

Problems

1. Over the temperature range from $-20°C$ to $+20°C$ laboratory data indicate that the surface tension of water varies inversely with temperature. The data can be accurately approximated by the formula $\sigma = AT + B$, where $A = -0.140$ ergs

$\mathrm{cm}^{-2}\,K^{-1}$ and $B = 113.6\,\mathrm{ergs\,cm}^{-2}$. Prove that, within the range from $-20°C$ to $+20°C$, the equilibrium vapor pressure over a pure water droplet increases with temperature if $r > 2B/L\rho_L$. Solve for the radius of a droplet at $273°K$ such that the change in equilibrium vapor pressure per degree of temperature increase just equals the change in equilibrium vapor pressure per micron of radius decrease.

2. Some condensation-nucleus counters currently in use consist of a cloud chamber whose relative humidity can be very precisely controlled, and a means of counting the number of cloud droplets formed within the chamber. Data are usually presented in the form of activity spectra, i.e. curves of the number of cloud droplets observed versus the supersaturation ratio.

Consider the case of an air sample containing an aerosol of NaCl with particle sizes ranging from r_0 to r_{max}, and with a distribution having the same slope as the Junge distribution. An experiment consists of drawing the sample into the chamber, of very gradually increasing the saturation ratio from a value less than 1 to a value considerably greater than 1, and of noting the number N of cloud droplets present as a function of supersaturation $(S - 1)$. Assume that droplets become visible only if they are activated, i.e. grow to their critical size r^*. Assuming further that $r_{max} \gg r_0$, show that N is related to supersaturation by

$$N \propto (S - 1)^2.$$

CHAPTER 6

DROPLET GROWTH BY CONDENSATION

Diffusional growth of a droplet

It was shown in Chapter 5 that a critical size r^* and saturation ratio S^* must be exceeded in order for a small solution droplet to become a cloud droplet. Before and after the droplet reaches the critical size, it grows by diffusion of water molecules from the vapor onto its surface. The rate of diffusional growth of a single droplet is analyzed in this section. Later we shall consider the more realistic case of many droplets simultaneously growing and competing for the available moisture.

The droplet has radius r and is located in a vapor field with the concentration of vapor molecules at distance R from the droplet center denoted by $n(R)$. The vapor field could as well be described in terms of the vapor density or absolute humidity $\rho_v(R)$, where $\rho_v = nm$ and m denotes the mass of one water molecule. Isotropy is assumed such that $n(R)$ or $\rho_v(R)$ does not depend on the direction outwards from the droplet. At any point in the vapor field the concentration of molecules is assumed to satisfy the diffusion equation

$$\frac{\partial n}{\partial t} = D\nabla^2 n, \tag{6.1}$$

where D is the molecular diffusion coefficient. Furthermore, steady-state or "stationary" conditions are assumed, so that $\partial n/\partial t = 0$. Then (6.1) becomes

$$\Delta^2 n(R) = 0 = \frac{1}{R^2}\frac{\partial}{\partial R}\left(R^2\frac{\partial n}{\partial R}\right), \tag{6.2}$$

with the general solution

$$n(R) = C_1 - C_2/R. \tag{6.3}$$

Boundary conditions are as follows:
as $R \to \infty$, $n \to n_0$, the "ambient" or undisturbed value of vapor concentration;
as $R \to r$, $n \to n_r$, the vapor concentration at the droplet's surface. The solution satisfying these conditions is

$$n(R) = n_0 - \frac{r}{R}(n_0 - n_r).$$ (6.4)

The flux of molecules onto the surface of the droplet is equal to $D(\partial n / \partial R)_{R=r}$. Consequently the rate of mass increase is given by

$$\frac{dM}{dt} = 4\pi r^2 D \left(\frac{\partial n}{\partial R}\right)_{R=r} m$$ (6.5)

where m is the mass of a water molecule.
    Utilizing (6.4), (6.5) becomes

$$\frac{dM}{dt} = 4\pi r D(n_0 - n_r)m.$$ (6.6)

In terms of the vapor density, this result may be written

$$\frac{dM}{dt} = 4\pi r D(\rho_v - \rho_{vr}),$$ (6.7)

where ρ_v is the ambient vapor density and ρ_{vr} is the vapor density at the droplet's surface.
    Associated with condensation is the release of latent heat, which tends to raise the droplet temperature above the ambient value. The diffusion of heat away from the droplet is given by an equation analogous to (6.7):

$$\frac{dQ}{dt} = 4\pi r K(T_r - T)$$ (6.8)

where T is ambient temperature, T_r is the temperature at the surface of the droplet, and K is the coefficient of thermal conductivity of air.
    Following (6.7) and (6.8), the rate of change of temperature at the droplet's surface is given by

$$\frac{4}{3}\pi r^3 \rho_L c \frac{dT_r}{dt} = L\frac{dM}{dt} - \frac{dQ}{dt}$$ (6.9)

where ρ_L is the density of water and c is its specific heat capacity. For the assumed steady-state growth process, $dT_r/dt = 0$, so that (6.9), when set equal to zero, leads to a balance condition that must be satisfied by the temperature and vapor density fields,

$$\frac{\rho_v - \rho_{vr}}{T_r - T} = \frac{K}{LD}. \tag{6.10}$$

In this equation the ambient conditions are described by p_v and T, and the ratio K/LD depends weakly on temperature and pressure. Ordinarily the drop temperature T_r and the vapor density at its surface ρ_{vr} are unknown. From (5.6) and the equation of state for water vapor, these quantities are related by

$$\rho_{vr} = e_s'\,(r)/R_vT_r = \left(1 + \frac{a}{r} - \frac{b}{r^3}\right)e_s\,(T_r)/R_vT_r, \tag{6.11}$$

where $e_s\,(T_r)$ is the equilibrium vapor pressure over a plane water surface at temperature T_r, and is given by the Clausius-Clapeyron equation. Equations (6.10) and (6.11) comprise a simultaneous system which can be solved numerically for T_r and ρ_{vr} in order to evaluate the rate of drop growth by condensation.

As an alternative to the numerical method of solution, Mason (1971) introduced a useful analytical approximation for calculating the rate of growth of a drop by condensation. In a field of saturated vapor, changes in vapor density are related to changes in temperature by

$$\frac{d\rho_v}{\rho_v} = \frac{L}{R_v}\frac{dT}{T^2} - \frac{dT}{T}. \tag{6.12}$$

Integrating this equation from temperature T_r to temperature T gives

$$\ln\frac{\rho_{vs}}{\rho_{vrs}} = (T - T_r)\left(\frac{L}{R_vT_rT} - \frac{1}{T_r}\right), \tag{6.13}$$

where the subscripts s indicate saturation vapor densities. Because $\rho_{vs}\,\rho_{vrs}$ is close to unity, (6.13) leads to the approximate relation

$$\frac{\rho_{vs} - \rho_{vrs}}{\rho_{vrs}} = \left(\frac{T - T_r}{T}\right)\left(\frac{L}{R_vT} - 1\right), \tag{6.14}$$

where the approximation has also been employed that $TT_r \approx T^2$. Substituting from (6.8) for $(T - T_r)$ gives

$$\frac{\rho_{sv} - \rho_{vrs}}{\rho_{vrs}} = \left(1 - \frac{L}{R_v T}\right)\left(\frac{L}{4\pi r K T}\right)\frac{dM}{dt}. \qquad (6.15)$$

From (6.7)

$$\frac{\rho_v - \rho_{vr}}{\rho_{vr}} = (4\pi r K T \ )^{-1}\frac{dM}{dt}. \qquad (6.16)$$

Subtracting (6.15) from (6.16), and assuming that $\rho_{vr} = \rho_{vrs}$, leads eventually to the approximate result

$$r\frac{dr}{dt} = \frac{S - 1}{\left[\left(\frac{L}{R_v T} - 1\right)\frac{L\rho_L}{KT} + \frac{\rho_L R_v T}{De_s(T)}\right]} \equiv \frac{S - 1}{\left[F_k + F_d\right]}, \qquad (6.17)$$

where $S = e/e_s(T)$ is the ambient saturation ratio. This form of the approximation neglects the solution and curvature effects on the drop's equilibrium vapor pressure. When these effects are included, the approximation becomes

$$r\frac{dr}{dt} = \frac{(S - 1) - \frac{a}{r} + \frac{b}{r^3}}{[F_k + F_d]} \qquad (6.18)$$

Here as in (6.17) F_k represents the thermodynamic term in the denominator that is associated with heat conduction; F_d is the term associated with vapor diffusion.

TABLE 6.1. *Values of Diffusion Coefficient D and Coefficient of Thermal Conductivity K. (From Smithsonian Meteorological Tables, 1958)*

T (deg C)	K (erg cm^{-1} sec^{-1} deg^{-1})	D (cm^2 sec^{-1})*
−20	2.28×10^3	0.197
−10	2.36×10^3	0.211
0	2.43×10^3	0.226
10	2.50×10^3	0.241
20	2.57×10^3	0.257
30	2.64×10^3	0.273
40	2.70×10^3	0.289

* Note: The tabulated values of D are for a pressure of 1000 mb. Since D is proportional to μ/ρ, it follows that D is inversely proportional to pressure for a given temperature. To obtain D for an arbitrary pressure p (mb), the tabulated values should therefore be multiplied by $(1000/p)$.

For typical values of a and b, it has been found that the droplet growth predicted by (6.18) is an excellent approximation to the growth rate obtained by simultaneous solution of (6.10) and 6.11).

The diffusion coefficient and thermal conductivity coefficient vary with temperature and are tabulated in Table 6.1. The latent heat L and equilibrium vapor pressure e_s also depend on temperature, and are tabulated in Table 2.1 of Chapter 2. As explained in the *Smithsonian Meteorological Tables* (R. J. List, ed.), K and D actually depend on the viscosity of the air. Specifically, K is proportional to the dynamic viscosity μ, and D is proportional to μ/ρ, with ρ the air density. The dynamic viscosity depends on temperature only, and is given approximately by the empirical formula

$$\mu(T) = 1.72 \times 10^{-4}\left(\frac{393}{T+120}\right)\left(\frac{T}{273}\right)^{3/2}$$

where T is in °K and μ is in g cm^{-1} sec^{-1}. This relation, combined with measured values of K and D at standard temperature and pressure, was used to generate the values given in the table. Figure 6.1 shows the dependence of the quantity $[F_k + F_d]$ (in sec cm^{-2}) on temperature and pressure.

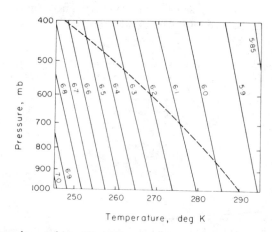

FIG. 6.1. Dependence of $\log_{10}[F_k + F_d]$ on temperature and pressure. Contours are drawn for the indicated values of $\log_{10}[F_k + F_d]$. The dashed line represents the pseudoadiabat corresponding to $\theta_w = 290$°K. In an unmixed, cloudy updraft, $[F_k + F_d]$ would vary as indicated along this curve.

After r becomes sufficiently large, a/r and b/r^3 are negligible compared to $(S - 1)$ and (6.17) is a good approximation. Then, if $(S - 1)/[F_k + F_d]$ is reasonably steady, the droplet radius increases with time according to

$$r(t) = \sqrt{r_0^2 + 2ct}, \qquad (6.19)$$

where
$$c = (S - 1)/[F_k + F_d]. \qquad (6.20)$$

The parabolic form of this growth law leads to a narrowing of the drop-size distribution as growth proceeds. For example, consider two droplets with initial radii of 1 μm and 10 μm, growing under the same conditions. From (6.19) it is easy to show that the larger droplet grows only to 14 μm during the same time the smaller droplet grows to 10 μm.

Solutions of (6.18) depend on temperature and the nature of the nucleus and must be obtained numerically. Table 6.2 illustrates the growth rate for nuclei of NaCl at $T = 273°K$ and $p = 900$ mb, with a supersaturation of 0.05%.

TABLE 6.2. *Rate of Growth of Droplets by Conden-sation (initial radius 0.75 μm) (From Mason, 1971)*

Nuclear mass (g)	10^{-14}	10^{-13}	10^{-12}
Radius (μm)	Time (sec) to grow from initial radius 0.75 μm		
1	2.4	0.15	0.013
2	130	7.0	0.61
5	1,000	320	62
10	2,700	1,800	870
20	8,500	7,400	5,900
30	17,500	16,000	14,500
50	44,500	43,500	41,500

The rate of evaporation of a droplet is also described by (6.18); in this case $S < 1$ and $(dr/dt) < 0$. By knowing the dependence of droplet fall speed on size, it is possible through (6.18) to solve for the distance through which a drop falls over the time required for it to evaporate completely. For drops smaller than about 50 μm in radius the terminal fall speed increases approximately with the square of the radius and the distance of fall for complete evaporation increases with the fourth power of radius. This approximation has been employed (and extended somewhat beyond its range of validity) to give the results in Table 6.3. A rapid increase of distance

TABLE 6.3. *Distance a Drop Falls before Evaporating, assuming Isothermal Atmosphere with $T = 280°K$, $S = 0.8$*

Initial radius	Distance fallen
1 μm	2 μm
3 μm	0.17 mm
10 μm	2.1 cm
30 μm	1.69 m
0.1 mm	208 m
0.15 mm	1.05 km

fallen with radius is evident, leading to a reasonable basis of discriminating between cloud droplets and raindrops. Raindrops are those large enough to reach the ground before evaporating. Cloud droplets are those small enough to evaporate soon after leaving the cloud. By convention the dividing line is drawn at $r = 0.1$ mm: drops larger than this size stand a good change of reaching the ground and are called raindrops. Actually the drops whose radii are near 0.1 mm are referred to as drizzle drops.

The growth of droplet populations

In natural clouds many droplets grow at the same time and compete for the available water vapor. When the droplets are large or numerous enough, the rate of depletion can exceed the rate of production of supersaturation, retarding or terminating the growth process.

Moisture is supplied by saturated air which is cooled in ascent. The amount of available moisture for cloud droplet growth at a given time is determined by the rate of supply and the rate of condensation. In general, the time rate of change of the saturation ratio may be written

$$\frac{dS}{dt} = P - C, \tag{6.21}$$

where P denotes a production term and C a condensation term. More specifically,

$$\frac{dS}{dt} = Q_1 \frac{dz}{dt} - Q_2 \frac{d\chi}{dt},\qquad (6.22)$$

where dz/dt is the vertical air velocity and $d\chi/dt$ is the rate of condensation, measured in grams of condensate per gram of air per unit time. Q_1 and Q_2 are thermodynamic variables given by

$$Q_1 = \frac{1}{T}\left[\frac{\varepsilon L g}{R' c_p T} - \frac{g}{R'}\right]\qquad (6.23)$$

$$Q_2 = \rho\left[\frac{R'T}{\varepsilon e_s} + \frac{\varepsilon L^2}{pTc_p}\right].\qquad (6.24)$$

Physically, $Q_1(dz/dt)$ is the increase of supersaturation due to cooling in adiabatic ascent; $Q_2(d\chi/dt)$ is the decrease in supersaturation due to the condensation of vapor on droplets. Q_1 and Q_2 are plotted in Fig. 6.2 as functions of temperature.

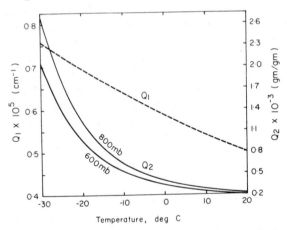

FIG. 6.2. The dependence of Q_1 and Q_2 on temperature. Q_2 depends also on pressure and is shown for 600 mb and 800 mb.

The derivation of (6.23) proceeds as follows. Assuming ascent with no condensation, (6.22) becomes

$$\frac{dS}{dt} = Q_1 \frac{dz}{dt}.\qquad (6.25)$$

But

$$S = e/e_s,$$

so

$$\frac{dS}{dt} = \left(e_s \frac{de}{dt} - e \frac{de_s}{dt} \right) \Big/ e_s^2. \qquad (6.26)$$

Introducing the mixing ratio w, and noting that w is constant when condensation is not occurring, you can show that

$$\frac{de}{dt} = -\frac{eg}{R'T}\frac{dz}{dt}. \qquad (6.27)$$

From the Clausius–Clapeyron equation,

$$\frac{de_s}{dt} = -\frac{Le_s}{R_v T^2}\frac{g}{c_p}\frac{dz}{dt}. \qquad (6.28)$$

Using (6.26)–(6.28) in (6.25) leads to the result (6.23) for Q_1. The derivation of (6.24) follows along similar lines.

On the basis of (6.18) for the growth rate and (6.22) for the saturation ratio, it is possible to start with an assumed or measured distribution of condensation nuclei, assume an updraft velocity, and calculate the subsequent evolution of the droplet spectrum. Such an analysis was first carried out by Howell (1949), whose results are discussed by Mason (1971) and Fletcher (1962) in their books on cloud physics. More recently Mordy (1959) made similar calculations using more up-to-date information on the nucleus spectrum. The largest nucleus size considered by Howell was about 10^{-12} g and Mordy's spectra extended to 10^{-9} g. Figure 6.3 (from Mordy) is for a "medium" nucleus concentration (one neither dense nor sparse) and an ascent rate of 15 cm/sec.

The solid lines indicate the growth of droplets spread over a wide range of initial nucleus sizes. The dashed line shows the variation with height of supersaturation. The characteristic narrowing of the size distribution—a consequence of the parabolic growth law—is pronounced. The supersaturation is seen to attain a maximum of about 0.5% at 40 m above cloud base. It is also at this altitude where the droplet growth curves indicate the fastest growth rate. The trajectory of the smallest droplet indicates that its critical supersaturation was approached but not reached before the saturation ratio started to decrease.

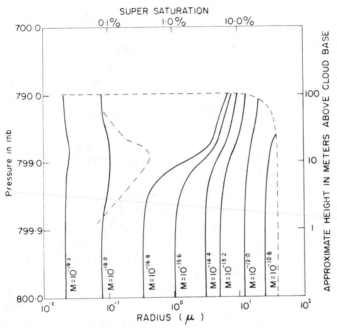

FIG. 6.3. Initial formation of cloud droplets and the variation of supersaturation above cloud base. (From Mordy, 1959.)

By comparing results for three assumed nucleus concentrations and three updraft velocities (15, 50, and 100 cm/sec), Mordy concluded that the number of cloud droplets produced depended both on the updraft and the nucleus concentration, with general agreement between calculated droplet concentrations and those observed in natural clouds. The calculated spectra, however, are narrower than most observed spectra and indicate too few large droplets. Twomey (1959), basing his analysis on empirical activity spectra of the form of (5.9), has carried out similar calculations, with results in accord with Mordy's findings.

A significant feature of all such calculations is that the supersaturation reaches its peak within about 100 m of cloud base, above which level it decreases and approaches an approximately constant value. Since supersaturation controls the number of condensation nuclei that are activated, the cloud droplet concentration is thus determined in the lowest cloud layer. The supersaturation decreases to its steady value when a balance is reached between the rate of condensation on the droplets which have formed and the updraft-produced rate of increase of supersaturation.

When the supersaturation may be regarded as steady there is a useful approximate method of determining the cloud droplet size distribution at any time, given its form at an earlier time. Let us consider a sample of cloudy air in which the drop-size distribution is characterized by the function $\nu_0(r_0)$, where $\nu_0(r_0)dr_0$ is the number of cloud droplets per unit mass of air with radii in the interval $(r_0, r_0 + dr_0)$. (The number of droplets per unit mass is related to the number $n(r)$ per unit volume of air by $n(r) = \rho\nu(r)$ where ρ is the air density.) At a later time t the droplets will have grown by condensation and the distribution will have changed as indicated schematically in Fig. 6.4. We assume that the droplets are large enough for the terms a/r and b/r^3 to be neglected in the growth equation (6.18). Then, since S is assumed constant, all drops grow according to the parabolic law (6.19). We assume further that none of the existing droplets leaves the sample, for example by evaporation or precipitation, and that no new droplets are added. Then the droplet mixing ratio, the total number per unit mass of air, remains constant. It follows from these assumptions that the drop-size distribution at time t is given in terms of the initial distribution by

$$\nu(r,t) = \nu_0(r_0)\frac{dr_0}{dr} = \frac{r}{\sqrt{r^2 - 2ct}}\,\nu_0(\sqrt{r^2 - 2ct}). \qquad (6.29)$$

Kovetz (1969) has derived a more general relation, which allows for the creation of new droplets and the fall of droplets relative to the air.

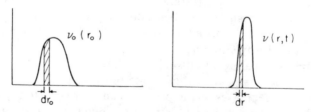

FIG. 6.4. Evolution of droplet spectrum by condensation (schematic).

While the theoretical results show an ever-narrowing droplet spectrum (as anticipated from the parabolic growth law), the spectra observed in clouds tend to broaden with time, and it is this broadening that may lead to the production of rain. Though there have been a number of attempts over the last decade to explain this inconsistency, none as yet has been generally accepted as the complete explanation. Some of these attempts are described in the next section.

Some modifications of the diffusional growth theory

(a) Growth of a single drop

(1) Borovikov *et al.* (1961) showed that the general solution of (6.1), without making the assumption of stationary growth, is

$$n(R) = n_0 - \frac{r}{R}(n_0 - n_r)\left[1 - \text{erf}\left(\frac{R-r}{2\sqrt{Dt}}\right)\right], \qquad (6.30)$$

where erf (x) denotes the error function. Comparing this result with the solution (6.4) for stationary growth, these authors demonstrated that the assumption of stationarity introduces only negligible errors in growth rate for droplets of typical size after an elapsed time of 10 μsec; hence the assumption of stationary growth is justified and cannot be responsible for the disparity between observed and calculated droplet spectra.

(2) As presented earlier, the diffusional growth equations neglect the fall velocity of the droplet relative to the vapor field. Such a velocity will introduce a "ventilation" effect that tends to increase the droplet growth rate. This effect may be taken into account by including a ventilation factor f, a function of the Reynolds and Prandtl numbers of the flow, in the growth equation (6.18). Fletcher (1962, p. 124) discussed the ventilation effect, showing that it is negligible for drops with $r < 10$ μm. With increasing droplet size it becomes more important; for $r \approx 80$ μm (at 10°C, 800 mb), the ventilation factor increases the quantity $r(dr/dt)$ by about 50%. This effect acts in the right direction to explain the disparity, but is too weak to account for much broadening in the important range from about 20 to 40 μm in radius.

(3) The growth equations that have been given are based on the

diffusion equation and do not specifically account for the molecular exchange processes at the liquid–vapor interface. It has long been recognized that this is a deficiency and a number of attempts have been made to correct the growth equations on the basis of kinetic theory. Fukuta and Walter (1970) have presented results of new calculations taking these effects into account, and have given a critical review of past work.

The problem is more complex from the kinetic theory approach, and two new concepts and parameters need to be introduced:

(a) The accommodation coefficient α, which describes the transfer of heat by molecules arriving at and leaving the interface between the liquid and vapor phases:

$$\alpha = \frac{T_2' - T_1}{T_2 - T_1},$$

where T_2' = temperature of vapor molecules leaving the surface of the liquid;
T_2 = temperature of the liquid;
T_1 = temperature of the vapor.

(b) Condensation coefficient β, defined as the fraction of molecules hitting the liquid surface which condense. It is found that polar liquids vaporize at rates slower than predicted using the equilibrium vapor pressure, and hence have lower values of β than nonpolar liquids.

In the kinetic treatment it is assumed that one kind of flow régime prevails within the distance of about a mean free path from the liquid surface and that another flow régime prevails outside this distance. The inner region is controlled by kinetic effects, the outer by ordinary diffusion. The inner region forms a barrier for heat and mass exchange, and the importance of this barrier is measured by the coefficients α and β.

For the growth equation Fukuta and Walter found (neglecting solution and curvature effects)

$$r\frac{dr}{dt} = \frac{S-1}{\left[\dfrac{L^2\rho_L}{KR_vT^2f(\alpha)} + \dfrac{R_vT\rho_L}{De_s(T)g(\beta)}\right]}, \qquad (6.31)$$

where the functions f and g depend on the kinetic parameters and the symbols otherwise are as in (6.17). Using the best available estimates for these functions, they found that after 20 sec of growth the mass of the droplet growing according to modified theory is about half the mass predicted by the more elementary diffusion theory. In (6.31) $g(\beta) < f(\alpha) \leq 1$, so that the diffusion term takes on relatively more importance than the heat conduction term in the modified theory. As growth continues beyond 20 sec the difference in mass predicted by the two theories becomes negligible. Thus the kinetic corrections are strongest felt by small droplets. Fukuta and Walter concluded that in situations where droplets remain small for a long time, or when considering a poly-dispersed system containing small or newly nucleated droplets, or a population of droplets with smaller β artificially induced, use of the revised theory is desirable.

Fitzgerald (1972) included the accommodation coefficient and condensation coefficient in a study of droplet growth by condensation. He noted that the instantaneous growth rates of droplets of 0.1, 1, and 10 μm radius are reduced, respectively, to 4, 33, and 82% of the growth rates computed neglecting these corrections. Since the kinetic factors tend to retard the growth of small droplets relative to larger ones, they act in the right direction to account for a broadening of the droplet spectrum with time. As was the case of ventilation effects, however, the kinetic effects are too small to account for many of the distributions actually observed.

(4) Basing their arguments on the statistical theory of turbulence, certain Soviet authors (e.g. Mazin, 1968) have concluded that local turbulence-induced pressure fluctuations within a cloud give rise to fluctuations in saturation ratio that can influence droplet growth. Each droplet experiences a time-varying supersaturation, and owing to the statistical nature of these variations, some droplets will grow more rapidly than the average. Consequently, after a 5–10 min growth period the dispersion of droplet sizes about their average might be increased by several microns due to this effect. The preliminary conclusions, as given by Mazin, suggest once again that this effect is not sufficient to account for observed droplet spectra.

(b) Growth of droplet populations

(1) In their analyses of the evolution of droplet spectra Mordy, Howell, and Twomey considered a discrete and confined parcel of

air containing a given distribution of condensation nuclei and ascending at a constant velocity. Fitzgerald (1972) used the same approach but had more complete observational data to compare with calculations. The input data consisted of the horizontally averaged updraft velocity measured (by airplane) several hundred feet below cloud base, the size distribution of NaCl nuclei inferred from a measured activity spectrum of condensation nuclei, and the actual cloud base temperature and pressure. Using the diffusional growth equation essentially in the form (6.31), Fitzgerald calculated the droplet spectrum at a height of a few hundred meters above cloud base and compared the results with observed spectra. Comparisons were made for five continental and two maritime clouds. Both the observed and computed spectra were quite monodisperse. Close agreement was found between observed and calculated spectra, as indicated by comparing the droplet concentrations, mean droplet diameters, and the standard deviation of diameters. The average values of the dispersion coefficient (the ratio of standard deviation to mean diameter) were 0.17 for the observed distributions and 0.12 for the computed distributions. An example of the results for one of the continental clouds is shown in Fig. 6.5.

Although Fitzgerald's study still shows a tendency for actual spectra to be somewhat broader than those predicted by theory, the disparity is not as great as had been thought earlier. The reason for the discrepancy thought to exist earlier was not a shortcoming in theory but inadequate observations. Fitzgerald measured the droplet distributions just above cloud base and found them to be almost as narrow as predicted by the diffusional growth theory. In earlier work the observations had not been confined to the lowest cloud level, and broader distributions were found.

Although the diffusional growth theory thus appears to explain adequately the early stages of droplet development there is not yet general agreement as to how droplet spectra, at some time after their initial formation, do broaden and extend over radius from several microns out to about 20 μm.

(2) In order to prove or disprove earlier speculation that turbulent mixing in the updraft would bring together parcels of different histories and thereby account for broad droplet spectra, Warner (1969b) numerically simulated a plume-type turbulent updraft and found that it gave spectra not significantly broader than those

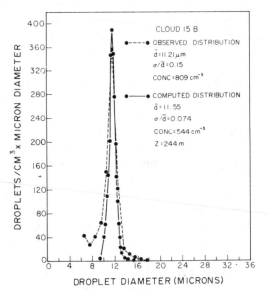

FIG. 6.5. Comparison of computed and measured cloud droplet spectrum at height of 244 m above cloud base. (From Fitzgerald, 1972.)

produced by a steady updraft. Some updraft structures, however, particularly those involving a general acceleration, are capable of broadening the spectrum by the continual activation of fresh nuclei, leading to a persistent and bimodal size distribution.

Warner included the condensation coefficient in his calculations. In order to account for the almost invariable observations of significant numbers of droplets having $r < 5$ μm at heights of 100 m or so above cloud base, he found it was necessary for this coefficient to be not too much larger than 0.05. But for droplets to grow beyond this size in reasonable times the coefficient cannot be much less than 0.03. (Note that this coefficient is β as introduced above, although Warner refers to it as the accommodation coefficient.) These estimates of the values of β are somewhat smaller than those employed by Fukuta and Walter.

In an approach different from Warner's, Mason and Jonas (1974) modeled turbulent mixing by considering successive buoyant bubbles ascending through the same volume of space. The sensitive parameters were found to be the environmental lapse rate and humidity and the rate of turbulent mixing between bubble and environment. The initial values of buoyancy and vertical velocity were less important. For reasonable values of the parameters, the model was found to account for the main features of the structure and constitution of non-precipitating cumulus clouds. Liquid water contents in the upper part of the simulated clouds were about 35% of the adiabatic value and peak values of supersaturation were about 0.4%. More important, a case based on input parameters representative of maritime conditions was shown to produce broad and bimodal droplet spectra in a time of about 30 min, with droplets as large as 25 μm. Interestingly enough, Mason and Jonas entirely neglected the kinetic corrections in the growth equation.

(3) Much earlier Mason (1952), considering stratiform clouds, noted that the updrafts are not irregular and assumed that the explanation for broad droplet distributions must be found elsewhere. One explanation offered was that some droplets simply stay in the cloud longer than others, and hence grow to larger sizes. This is a plausible reason for some of the spectral broadening, and seems to be in accord to some extent with the recent findings of Mason and Jonas.

Problems

1. A raindrop falls from a cloud into an isothermal, unsaturated layer of air below. As the drop evaporates the rate at which it loses heat by evaporation equals the rate at which it gains heat by conduction from the warmer air. Using the data in Table 6.1, show that the temperature of the drop approximately equals the wet-bulb temperature of the air.

2.* A towering cumulus congestus cloud has its base at 820 mb and 14°C. At 770 mb the cloud liquid water content is 1 g/kg and the drop spectrum is of Gaussian shape, centered at a radius of 8 μm and with a dispersion $\sigma/\bar{r}$ of 0.15. Assume that air ascends in the cloud pseudoadiabatically with no entrainment. Assuming droplet growth by condensation only, find the form of the drop spectrum at 500 mb. Assume that the amount of water condensed is always χ, the adiabatic liquid water content.

Assuming further a constant vertical velocity of 15 m/sec from 770 to 500 mb, estimate the average supersaturation of the ascending air.

In your calculations neglect a/r and b/r^3 in the droplet growth equation. A tephigram is useful in solving this problem.

---

* This problem makes rather severe computational demands and is best approached with a digital computer.

CHAPTER 7

INITIATION OF RAIN IN NONFREEZING CLOUDS

MOST of the world's precipitation falls to the ground as rain, much of which is produced by clouds whose tops do not extend to temperatures colder than 0°C. The mechanism responsible for precipitation in these "warm" clouds is coalescence among cloud droplets. By far the dominant precipitation-forming process in the tropics, coalescence also plays a role in midlatitude cumulus clouds whose tops may extend to subfreezing temperatures.

In this and subsequent chapters it is assumed that the reader is acquainted with the major categories of cloud classification. A number of cloud atlases are available with excellent illustrations of the various cloud types, for example Scorer and Wexler's (1963) *A Colour Guide to Clouds.*

Microphysical properties of clouds

Cloud droplet spectra may be characterized by the function $n(r)$, with the property that $n(r)dr$ is the number of droplets per unit volume with radii in the interval $(r, r + dr)$. (See Fig. 6.5 of the preceding chapter for an example.) In general the distribution will vary with position in a cloud and with time at any one location. In addition to the real variability there is statistical variability arising from the randomness of droplet locations. To suppress the statistical variability any measurement of $n(r)$ should be made in a cloud volume large enough to contain a relatively large number of droplets in each size category.

The droplet concentration or number density N is the total number of droplets per unit volume, and equals the integral of $n(r)$ over all droplet sizes present. It also is a variable quantity and subject to sampling limitations. In continental cumulus clouds a typical value of N is $200 \, \text{cm}^{-3}$; in Hawaiian orographic clouds, an extreme case, about $10 \, \text{cm}^{-3}$. These are also the concentrations predicted by diffusional growth theory applied to populations of ascending condensation nuclei.

The cloud liquid water content M is the mass of condensed water per unit volume of air, defined by

$$M = \tfrac{4}{3}\pi\rho_L \int r^3 n(r)\,dr.$$

In non-precipitating cumulus clouds a typical value of M is $0.5 \, \text{g/m}^3$, with peak values of about $1 \, \text{g/m}^3$. In stratus clouds the values tend to be smaller. In cumulonimbus clouds M can exceed $5 \, \text{g/m}^3$. The upper limit of M is approximately M_a, the adiabatic liquid water content from parcel theory.

Table 7.1, extracted in part from material given by Fletcher (1962), summarizes typical cloud characteristics.

TABLE 7.1. *Characteristics of Typical Clouds*

Cloud type	Droplet number density (cm^{-3})	Median droplet radius (μm)	Liquid water content* (g/m^3)	Thickness for 20% precip. probability (km)
Hawaiian Orographic	10	20	0.5	†
Maritime Cumulus	50	15	0.5	2.5
Continental Cumulus	200	5	0.3–3.0	6

* These figures are quite variable, depending upon extent of cloud development.
† Although probability estimates are not available, clouds of this type no thicker than 2 km frequently produce light rain.

Figure 7.1, from aufm Kampe and Weickmann (1957), shows average values of droplet spectra from various cloud types, plotted on semilogarithmic coordinates. Only the curves for stratocumulus and altocumulus clouds are seen to approach the one in Fig. 6.5 in narrowness. Cumulus congestus, and especially cumulonimbus, have relatively high droplet concentrations and appreciable numbers of drops larger than 50 μm.

Warner (1969a) gave examples of strongly bimodal droplet spectra

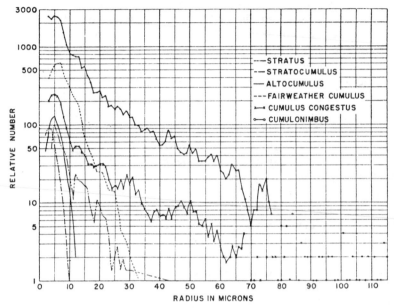

FIG. 7.1. Average cloud droplet spectra for different cloud types. (From aufm Kampe and Weickmann, 1957.)

(in several cases with peaks at about 6 μm and 12 μm radius) and cautioned that some of the unimodal and positively skewed distributions reported in the literature are the consequence of averaging a large number of individual spectra, some of which may be much different from the resultant average.

Droplet growth by collision and coalescence

Collisions may occur through differential response of the droplets to gravitational, electrical, or aerodynamic forces. Gravitational effects predominate in clouds: large droplets fall faster than smaller ones, overtaking and capturing a fraction of those lying in their paths. The electrical and turbulent fields required to produce a comparable number of collisions are much stronger than those thought usually to exist, although the intense electric fields in thunderstorms may create significant local effects. As a drop falls it will collide with only a fraction of the droplets in its path because some will be swept aside in the airstream around the drop. The ratio of the actual number of collisions to the number for complete geometric sweep-out is called the collision efficiency, and depends primarily on the size of the collector drop and the sizes of the collected droplets.

Collision does not guarantee coalescence. When a pair of drops collide several types of interaction are possible: (1) they may bounce apart; (2) they may coalesce and remain permanently united; (3) they may coalesce temporarily and separate, apparently retaining their initial identities; (4) they may coalesce temporarily and then break into a number of small drops. The type of interaction depends upon the drop sizes and collision trajectories, and is also influenced by the existing electrical forces and other factors. For sizes smaller than 100 μm in radius, the important interactions are (1) and (2) in the preceding list. The ratio of the number of coalescences to the number of collisions is called the coalescence efficiency. The growth of a drop by the collision-coalescence process is governed by the *collection efficiency*, which is the product of collision efficiency and coalescence efficiency. Laboratory studies of small colliding drop-lets indicate that the coalescence efficiency is close to unity if the droplets are charged or an electrical field is present. Because weak fields and charges exist in natural clouds, theoretical studies of droplet growth by collision-coalescence usually make the assumption that the collection efficiency equals the collision efficiency. The problem of explaining the initial development of rain then reduces to one of determining collision rates among a population of droplets.

(a) Droplet terminal fall speed

The drag force exerted on a sphere of radius r by a viscous fluid is given by

$$F_R = \frac{\pi}{2}r^2u^2\rho C_D, \tag{7.1}$$

where u is the velocity of the sphere relative to the fluid, ρ is the fluid density, and C_D is the drag coefficient characterizing the flow. In terms of the Reynolds number $Re = 2\rho ur/\mu$, with μ the dynamic viscosity of the fluid, (7.1) may be written in the form

$$F_R = 6\pi\mu ru(C_D Re/24). \tag{7.2}$$

The gravitational force on the sphere is given by

$$F_G = \tfrac{4}{3}\pi r^3 g(\rho_L - \rho),$$

where ρ_L is the density of the sphere. For the case of a water drop falling through air $\rho_L \gg \rho$ and

$$F_G = \tfrac{4}{3}\pi r^3 g\rho_L \qquad (7.3)$$

to good approximation. When $F_G = F_R$ the drop falls relative to the air at its terminal fall speed. For this equilibrium situation

$$u^2 = \frac{8}{3}\frac{rg\rho_L}{\rho C_D}$$

or

$$u = \frac{2}{9}\frac{r^2 g\rho_L}{(C_D Re/24)\mu}. \qquad (7.4)$$

For very small Reynolds numbers the Stokes solution[*] to the flow field around a sphere shows that $(C_D Re/24) = 1$. In this case, (7.4) reduces to

$$u = \frac{2}{9}\frac{r^2 g\rho_L}{\mu} = k_1 r^2, \qquad (7.5)$$

with $k_1 \approx 1.19 \times 10^6\,\text{cm}^{-1}\,\text{sec}^{-1}$. This quadratic dependence of fall speed on size is called Stokes' Law and applies to cloud droplets up to about 40 μm radius.

Experiments with spheres indicate that for sufficiently high Reynolds numbers C_D becomes independent of Re and has a value of about 0.45. Using this value in (7.4) leads to

$$u = k_2 r^{1/2}, \qquad (7.6)$$

where

$$k_2 = 2.2 \times 10^3 \left(\frac{\rho_0}{\rho}\right)^{1/2} \text{cm}^{1/2}\,\text{sec}^{-1}. \qquad (7.7)$$

In (7.7) ρ is the air density and ρ_0 is a reference density of $1.20 \times 10^{-3}\,\text{g/cm}^3$, corresponding to dry air at 1013 mb and 20°C. Raindrops have high Reynolds numbers but are not perfectly spherical. Consequently, though often a useful approximation, (7.6) describes the fall speed of raindrops reasonably well only over a limited range of size.

---

[*] See, for example, Lamb (1945), pp. 598–599.

The most accurate observational data available on raindrop fall speed are those given by Gunn and Kinzer (1949), reproduced here in Table 7.2. These data were obtained at sea-level conditions, 1013 mb and 20°C. Owing to the reduced air density, a droplet of given size will tend to fall faster aloft than at sea level, approximately in accordance with the square-root law of (7.7). Beard (1976) developed empirical formulas which fit the data accurately and which can also be corrected for temperature and pressure. At the surface, raindrops which have attained the largest possible size before breakup fall no faster than about 9 m/sec. Under typical conditions at 500 mb the upper limit is about 13 m/sec. Beard's formulas apply to three separate ranges of diameter and are rather complicated.

TABLE 7.2. *Terminal Fall Speed as a Function of Drop Size (equivalent spherical diameter) (From Gunn and Kinzer, 1949)*

Diam. (mm)	Fall speed (m/sec)	Diam. (mm)	Fall speed (m/sec)
0.1	0.27	2.6	7.57
0.2	0.72	2.8	7.82
0.3	1.17	3.0	8.06
0.4	1.62	3.2	8.26
0.5	2.06	3.4	8.44
0.6	2.47	3.6	8.60
0.7	2.87	3.8	8.72
0.8	3.27	4.0	8.83
0.9	3.67	4.2	8.92
1.0	4.03	4.4	8.98
1.2	4.64	4.6	9.03
1.4	5.17	4.8	9.07
1.6	5.65	5.0	9.09
1.8	6.09	5.2	9.12
2.0	6.49	5.4	9.14
2.2	6.90	5.6	9.16
2.4	7.27	5.8	9.17

It can be determined from the data that (7.6) provides a reasonable approximation to the fall speed in the radius interval 0.6 mm $< r <$ 2 mm, but with $k_2 \approx 2.01 \times 10^3$ cm$^{1/2}$ sec^{-1}. In the intermediate size range, between the Stokes' Law region and the square-root law, an

approximate formula for fall speed is

$$u = k_3 r, \qquad 40\mu\text{m} < r < 0.6\,\text{mm} \tag{7.8}$$

with $k_3 = 8 \times 10^3 \,\text{sec}^{-1}$.

(b) Collision efficiency

A drop of radius R is pictured in Fig. 7.2 overtaking a droplet of radius r. If the droplet had zero inertia it would be swept aside by the stream flow around the larger drop and a collision would not occur. Whether a collision does in fact occur depends on the relative importance of the inertial force and the aerodynamic force, and the separation x between drop centers, called the impact parameter. For given values of r and R there is a critical value x_0 of the impact parameter within which a collision is certain to occur and outside of which the droplet will be deflected out of the path of the drop. Recent calculations summarized by Mason (1971) have now accurately established the values of x_0 over a wide range of spherical drop sizes. Results are presented in the form of collision efficiencies, defined by

$$E(R,r) = \frac{x_0^{\,2}}{(R + r)^2}. \tag{7.9}$$

So defined, the collision efficiency is equal to the fraction of those droplets with radius r in the path swept out by the collector drop that actually collide with it. Alternately, E may be interpreted as the probability that a collision will occur with a droplet located at random in the swept volume. Clearly $E \leqslant 1$.

FIG. 7.2. Collision geometry.

Figure 7.3 presents collision efficiencies for small collector drops as obtained from three sets of theoretical calculations. The solid curves of Hocking (1959) were accepted as accurate for about a decade and figure prominently in the cloud physics texts of Fletcher and Byers. These results have now been superseded by two more recent calculations, which are seen to agree fairly well with each other. One of the most significant differences between Hocking's original results and the new calculations concerns the "18-μm cutoff". The early work indicated that collision was impossible unless the larger droplet exceeded 18 μm in radius. The new results show that collisions can still occur for collector drops at least as

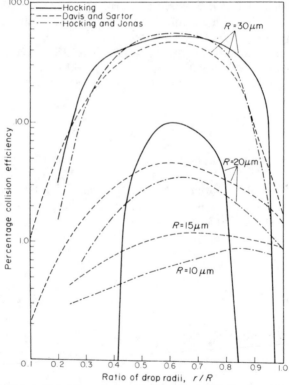

FIG. 7.3. Collision efficiencies calculated by Hocking (1959), Davis and Sartor (1967), and Hocking and Jonas (1970). (From Mason, 1971.)

small as 10 μm, though with small efficiency. The new calculations also indicate small but non-zero collision efficiency for droplets of the same size, unlike the original findings of Hocking.

These results for relatively small collector drops, as well as the best available data for larger collector drops, are summarized in Table 7.3. The entries in this table were used to prepare Fig. 7.4, the field of collision efficiency as a function of R and r. The figure shows that E is generally an increasing function of R and r, but for R greater than about 80 μm E depends largely on r.

TABLE 7.3. *Collision Efficiency E for Drops of Radius R colliding with Droplets of Radius r at 0°C and 900 mb.* (From Mason, 1971)

$R(\mu m)$	$r(\mu m)$							
	2	3	4	6	8	10	15	20
15		0.003	0.004	0.006	0.010	0.012	0.007	—
20	0.002	0.002	0.004	0.007	0.015	0.023	0.026	—
25	—	—	—	0.010	0.026	0.054	0.130	0.06
30	*	*	*	0.016	0.058	0.17	0.485	0.54
40	*	*	—	0.19	0.35	0.45	0.60	0.65
60	*	*	0.05	0.22	0.42	0.56	0.73	0.80
80	—	—	0.18	0.35	0.50	0.62	0.78	0.85
100	0.03	0.07	0.17	0.41	0.58	0.69	0.82	0.88
150	0.07	0.13	0.27	0.48	0.65	0.73	0.84	0.91
200	0.10	0.20	0.34	0.58	0.70	0.78	0.88	0.92
300	0.15	0.31	0.44	0.65	0.75	0.83	0.96	0.91
400	0.17	0.37	0.50	0.70	0.81	0.87	0.93	0.96
600	0.17	0.40	0.54	0.72	0.83	0.88	0.94	0.98
1000	0.15	0.37	0.52	0.74	0.82	0.88	0.94	0.98
1400	0.11	0.34	0.49	0.71	0.83	0.88	0.94	0.95
1800	0.08	0.29	0.45	0.68	0.80	0.86	0.96	0.94
2400	0.04	0.22	0.39	0.62	0.75	0.83	0.92	0.96
3000	·0.02	0.16	0.33	0.55	0.71	0.81	0.90	0.94

* E takes values of order 1% in this range but no accurate values have been computed.

Some of the papers on collision efficiency refer to a quantity called the linear collision efficiency, defined by

$$y_c = x_0/R. \tag{7.10}$$

From (7.9) and (7.10),

$$E = y_c^2/(1+p)^2, \tag{7.11}$$

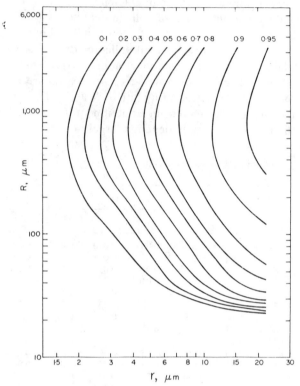

FIG. 7.4. Field of $E(R,r)$ based on data in Table 7.3.

where $p = r/R$. Alternatively the efficiency is sometimes defined by

$$E' = x_0^2/R^2.$$

Since $E' = E(1 + p)^2$, it is clear that E' can take on values greater than unity.

(c) Growth equations

Suppose a drop of radius R is falling at terminal speed through a population of smaller droplets. During unit time it sweeps out droplets of radius r from a volume given by

$$\pi(R + r)^2 [u(R) - u(r)]$$

where u denotes terminal fall speed. Thus the average number of droplets with radii between r and $r + dr$ collected in unit time is given by

$$\pi(R + r)^2 [u(R) - u(r)] n(r) E(R,r) dr$$

where $E(R,r)$ denotes the *collection* efficiency, which equals the product of collision efficiency times coalescence efficiency. When the drops are all smaller than about 100 μm it is usually assumed that the coalescence efficiency is unity, so that the collection efficiency is identical to the collision efficiency.

The total rate of increase in volume of the collector drop is obtained by integrating over all droplet sizes:

$$\frac{dV}{dt} = \int_0^R \pi(R + r)^2 \frac{4}{3} \pi r^3 E(R,r) n(r) [u(R) - u(r)] \, dr. \qquad (7.12)$$

In terms of drop radius

$$\frac{dR}{dt} = \frac{\pi}{3} \int_0^R \left(\frac{R + r}{R}\right)^2 [u(R) - u(r)] n(r) r^3 E(R,r) dr. \qquad (7.13)$$

The change of drop size with altitude may be obtained from

$$\frac{dR}{dz} = \frac{dR}{dt} \frac{dt}{dz} = \frac{dR}{dt} \frac{1}{w - u(R)}. \qquad (7.14)$$

(7.13) is general in that it allows for the sizes and fall speeds of the collected droplets. If these droplets are much smaller than the collector drop, then an approximation to (7.13) follows by setting $u(r) \approx 0$ and $R + r \approx R$. Thus

$$\frac{dR}{dt} = \frac{\bar{E}M}{4\rho_L} u(R). \qquad (7.15)$$

where $\bar{E}$ is an effective average value of collection efficiency for the droplet population and M is the cloud liquid water content. From (7.14), the change of radius with altitude is then given approximately by

$$\frac{dR}{dz} = \frac{\bar{E}M}{4\rho_L} \frac{u(R)}{w - u(R)}.$$ (7.16)

If the updraft is negligibly small, then

$$\frac{dR}{dz} = -\frac{\bar{E}M}{4\rho_L}.$$

The Bowen model

The simplified growth equation (7.16), taken in connection with (6.17) for diffusional growth, was used by Bowen in 1950 in an assessment of rain development in warm clouds, as described by Fletcher (1962). A cloud of uniformly-sized droplets was assumed to be ascending in a uniform updraft, and a slightly larger drop, presumably formed by the chance coalescence of two droplets, was assumed to be present. An example of the results of this model is shown in Fig. 7.5.

Initially the drop is carried with the droplets in the updraft. After a while, however, it begins to lag behind because its size increases. Growth then proceeds more rapidly and after a further period of time the drop becomes large enough to descend through the updraft. Growth continues as the drop falls through the cloud and emerges as a raindrop.

Important parameters in Bowen's model are the updraft velocity and the cloud water content. With increasing updraft, the drop ascends to a higher level before beginning its descent, and emerges from cloud base with a larger size. For a given updraft, drops grow larger but have a lower trajectory as the liquid water content increases.

Bowen assumed $E = 1$ in his calculations since information was lacking on the collision efficiencies of small drops. It is now known that for the small sizes he considered—$r = 10 \ \mu m$ and $R = 12.6 \ \mu m$ initially—the collision efficiency is too small for any growth to occur.

To illustrate the effect of the updraft in Bowen's model, Figs. 7.6 and 7.7 have been constructed using the data on collision efficiency from Table 7.3. As in Bowen's calculations, $r = 10 \ \mu m$ and $M =$

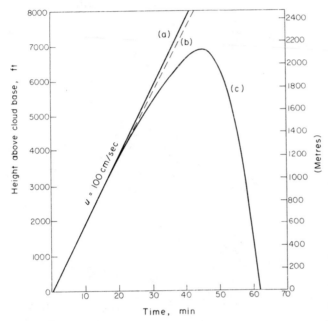

FIG. 7.5. Bowen's calculated trajectories of (a) the air, (b) cloud droplets, initially 10 μm in radius, and (c) drops which have initially twice the mass of the cloud droplets. Updraft speed 1 m/sec, cloud water content $M = 1$ g/m^3. (From Fletcher, 1962.)

1 g/m^3, but to insure some growth of the collector drop its initial radius was set at 20 μm. Figure 7.6 shows the trajectories for updrafts of 0.5 and 1 m/sec; Fig. 7.7 shows the drop diameter as a function of height. Droplet growth by condensation was neglected.

Calculations of this sort establish associations among updraft, cloud height, time for rain production, and the size of drops produced, that are in qualitative accord with observations. The most serious discrepancy between model predictions and observations lies in the time requirement. While calculations indicate that about an hour is required to produce millimeter-sized drops, observations show that such drops can be formed in less than half this time. Much of the subsequent work on rain development has been directed to the question of how the growth time can be reduced.

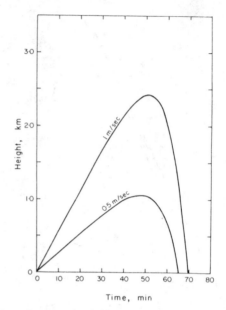

FIG. 7.6. Drop trajectories calculated for the collision efficiencies of Table 7.3 and Fig. 7.4, assuming a coalescence efficiency of unity. Initial drop radius 20 μm. Cloud water content 1 g/m^3; all cloud droplets of 10 μm radius.

Statistical growth: the Telford model

Even in a well-mixed cloud with the same average droplet concentration throughout, there will be local variations in concentration. In particular, if $\bar{n}$ denotes the average concentration of droplets in a given size interval, then the number m of such droplets in a volume V obeys the Poisson probability law,

$$p(m) = e^{-\bar{n}V} \frac{(\bar{n}V)^m}{m!}. \tag{7.17}$$

The growth equations in section (c) above do not take into account the statistical fluctuations and therefore apply only to average droplet growth.

It is not the average growth that figures into the development of rain, however. Some statistically "fortunate" drops fall through regions of locally high droplet concentration, experiencing more

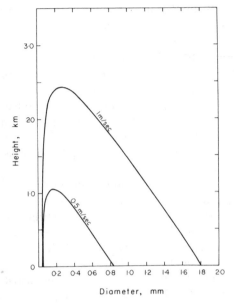

FIG. 7.7. Drop diameters for the trajectories of Fig. 7.6.

than the average number of collisions early in their development, and are subsequently in a favored position to continue to grow relatively rapidly. Rain is produced when only one such drop out of 10^5 or 10^6 gets an initial head start on its neighbors and then grows to raindrop size by gravitational coalescence. The time required for this to occur may be considerably shorter than the time required for an average ·droplet to reach raindrop size. The growth equations above view the coalescence process as being smooth and continuous: for any increment of time Δt, no matter how small, (7.15) for example can be used to solve for the increase in drop radius. In fact the drop grows not by a continuous process but by discrete collision and capture events.

Telford (1955) recognized these shortcomings of the "continuous-growth" equations and formulated a coalescence model taking into account the discrete nature of the growth process and the statistical fluctuations of droplet concentration. He assumed that the collected droplets were all the same size (10 μm radius in the most relevant

case considered) and that the collector drops have twice the volume
(12.6 μm radius). From the results Telford concluded that the
statistical-discrete capture process is crucial in the early stages of
rain formation. From the initial bimodal drop-size distribution,
reasonable raindrop spectra evolved over periods of a few tens of
minutes. Moreover, in one fairly representative case, he found that
100 of the 12.6 μm drops per cubic meter would experience their
first 10 coalescences after a time of only about 5 min. Compared to
this, the time required for 100 drops/m^3 to undergo 10 collisions
under the same conditions but assuming continuous growth would
be 33 min.

In his analysis, Telford derived first the probability distribution of
the time required for the collector drops to experience a given
number of collisions. Examples of such distributions are sketched
schematically in Fig. 7.8. Here Mdt represents the proportion of
drops that experience n captures ($n = 20, 30$ in this illustration) in
the time between t and $t + dt$. The most likely times required for n
captures correspond to the peak values of the distributions and are
indicated by t_{20} and t_{30} in this example. The most likely time for a
given number of collisions is very nearly equal to the time predicted
by the continuous-growth equations. The distributions are slightly
skewed, however, with the result that the average time for n
collisions exceeds somewhat the time for continuous growth to the
same size. The drops that grow faster than the average and can
account for the relatively rapid development of rain are those in the
left-hand tails of the distributions. Telford found that after about 20

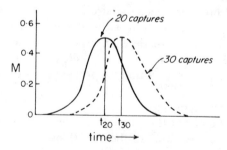

FIG. 7.8. Distribution curves of time required for 20 and 30 collisions of a collector
drop (schematic).

collisions the *shapes* of these distributions remained unchanged; they only shifted out along the time axis with the position of the maximum predictable by continuous-growth theory. He concluded from this behavior that the statistical effects were important only for about the first 20 collisions by which time the distribution was established, and after which the continuous-growth equations could be applied.

Because reliable information on the collision efficiencies of small drops was not available at the time, Telford assumed $E = 1$ in his calculations. This assumption made it possible to solve the equations analytically, but it is now known to be greatly in error for the small drop sizes considered. Using basically the same approach as Telford, Robertson (1974) has estimated the importance of statistical effects in coalescence growth, employing the best available data on collision efficiency. Owing to the complex form of $E(R,r)$ (see Fig. 7.4) analytical solutions are not possible and Robertson used a Monte Carlo procedure to simulate the collisions of a collector drop. Like Telford, Robertson found that the time distributions approach a limiting form as the number of collisions increases. Several simulations were carried out, with cloud droplet sizes r ranging between 8 and 14 μm in radius, and collector drops with initial sizes $R(0)$ ranging from 20 to 40 μm. An example of Robertson's results is shown in Fig. 7.9.

As n increases, the standard deviations increase and approach a limiting value. From Fig. 7.9 and results for the other cloud droplet sizes, it was found that the limiting value is achieved after a sufficient number of collisions of the collector drop: approximately 6 for $R(0) = 20$ μm, 40 for $R(0) = 30$ μm, and 100 for $R(0) = 40$ μm. The limiting values of σ decrease with increasing $R(0)$, and are negligibly small for drops greater than 40 μm in radius, implying that the continuous growth equations may be used. For $R(0)$ less than 20 μm, the collision efficiencies are too small to allow development of drizzle drops in times less than several hours, even though the statistical deviations from average drop behavior are large.

Statistical growth: the stochastic coalescence equation

In the analyses of Bowen, Telford, and Robertson droplet growth starts with a population consisting of distinct collected droplets and

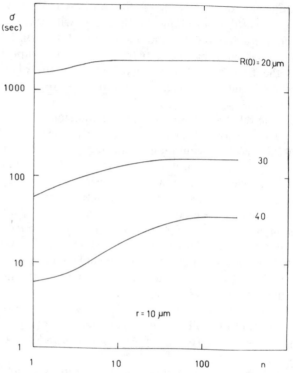

FIG. 7.9. The standard deviation σ of time required to make n captures, for $r = 10$ μm and $R(0) = 20$, 30, and 40 μm. Cloud liquid water content $M = 1$ g/m³. (From Robertson, 1974.)

collector drops. What must happen in nature is that a continuous spectrum of droplet sizes, formed by the condensation-diffusion process, evolves by random collisions (at first very rare) and thereby extends itself in the direction of increasing drop size. To be understood, the rain-forming process must therefore be viewed as the evolution of an entire droplet spectrum rather than as the growth of a subset of drops, assumed to be the "collectors" from the start.

To derive the differential equation describing the development in time of the droplet spectrum, we begin by defining the *coagulation coefficient* that describes the likelihood that a drop of radius R will overtake and collide with a droplet of radius r. Suppose these two drops are contained in a unit volume of air. In a unit time, the larger drop sweeps out a volume given by $\pi R^2 u(R)$ and the droplet sweeps out volume $\pi r^2 u(r)$. If the drop is to capture the droplet in unit time, neglecting aerodynamic effects, we see that they cannot be far apart initially and that, in fact, they must be contained in a common volume given by $\pi(R + r)^2[u(R) - u(r)]$. Allowing for aerodynamic effects reduces the size of this common volume to

$$K(R,r) = \pi(R + r)^2 |u(R) - u(r)| E(R,r), \qquad (7.18)$$

where $E(R,r)$ is the collision efficiency. The quantity $K(R,r)$ is called the coagulation coefficient. The continuous growth equations (7.12) and (7.13) could have been formulated in terms of K. In the stochastic equations this coefficient is interpreted as the *probability* that a drop of radius R collects a droplet of radius r in unit time, given that both are present with unit concentration.

While it is natural to define the coagulation coefficient in terms of drop radii, the stochastic equations turn out to be simpler in appearance if it is transformed to a function of drop volumes. Letting V and v denote the volumes corresponding to radii R and r, we have for the coagulation coefficient

$$H(V,v) = K\left[\left(\frac{3V}{4\pi}\right)^{1/3}, \left(\frac{3v}{4\pi}\right)^{1/3}\right]. \qquad (7.19)$$

Thus $H(V,v)$ describes the probability that a drop of volume V will collect a droplet of volume v.

Now we suppose that the drop spectrum is characterized by $n(v)$ such that $n(v)dv$ is the average number of drops per unit volume of space whose volumes are between v and $v + dv$. The total number of coalescences per unit time experienced by drops within this size interval is

$$n(v)dv \int_0^\infty H(V,v)n(V)dV.$$

This integration accounts for all possible captures of the drops in dv by larger drops ($v < V < \infty$), as well as all captures of smaller droplets by the drops in $dv (0 < V \leq v)$. These coalescence events *reduce* the number of drops in dv. But the number of drops in this size interval is *increased* by coalescences between all pairs of smaller drops whose volumes sum to v. This rate of increase is given by

$$\frac{1}{2} dv \int_0^v H(\delta,u)n(\delta)n(u)du,$$

where $\delta = v - u$. The $\frac{1}{2}$ factor is necessary to prevent any particular capture combination from being counted twice.

Taking into account both effects, we have for the rate of change of drop concentration in the size interval dv

$$\frac{\partial}{\partial t} n(v)dv = \frac{1}{2} dv \int_0^v H(\delta,u)n(\delta)n(u)du$$

$$- n(v)dv \int_0^\infty H(V,v)n(V)dV. \qquad (7.20)$$

Variously called the kinetic equation, coagulation equation, or stochastic coalescence equation, this result dates back to Smoluchowski in 1916 according to Drake (1972a) in a comprehensive review. It was first employed in the analysis of rain formation more than twenty years ago by Melzak and Hitschfeld (1953). Although the formulation of the problem was correct, this early work was impaired for two reasons: the collision efficiencies for small drops, on which the results crucially depend, were not accurately known; moreover, since a high speed computer was not available, the integrations had to be done by hand.

It was a surprisingly long time after computers became generally available that this approach was again undertaken—in this instance by Twomey (1964, 1966). Twomey, motivated by Telford's work described above, apparently rederived the stochastic equation, claiming that this approach embodied, in a more general form, the idea of the statistically "fortunate" drops. It was argued from the first (Warshaw, 1967) that this interpretation of (7.20) is incorrect, partly because the equation is based on the mean drop-size distribution $n(v)$. Controversy has continued on whether the equation actually embodies the statistical effects ascribed to it. Theoretical doubts were aggravated by the fact that early results on stochastic coalescence were often in disagreement, owing to numerical errors that arise in the numerical integration of (7.20). By experimenting with a number of combinations of integration and interpolation schemes, Reinhardt (1972) has established numerical procedures for integrating (7.20) which are very accurate, at least so long as the initial distribution is not too peaked. On the theoretical side there have been contributions by Scott (1968, 1972), Long (1971), Drake

(1972b), and most recently Gillespie (1972, 1975). Although there remain some rather subtle unresolved points, it appears from Gillespie's analysis that (7.20) does include the intended statistical effects.

The solution of (7.20) is $n(v,t)$, the droplet spectrum at time t which evolves by coalescence from a given initial distribution $n(v,0)$. Since in nature the growth is stochastic, it must be recognized that a given distribution $n(v,0)$ leads to an entire family of solutions $n(v,t)$, one for each "realization" of the coalescence process. The deterministic solution provided by (7.20) corresponds to the average value of $n(v,t)$ over many such realizations. Gillespie showed that, subject to reasonable assumptions, the statistical fluctuations about $n(v,t)$ are Poisson-distributed with parameter n. Consequently the statistical dispersion of $n(v,t)$ diminishes as $1/\sqrt{n}$.

An example of the results obtained by integrating (7.20) is shown in Fig. 7.10. The initial droplet distribution was assumed to be Gaussian in form, with the ratio $\sigma/\bar{r}$ of standard deviation to mean droplet radius equal to 0.15, and with a liquid water content of 1 g/m³. The stochastic equation was integrated with different assumed values of the collection efficiency to give these results. As expected, the distribution evolves fastest for geometric sweepout

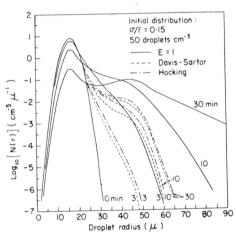

FIG. 7.10. Droplet spectrum at 3, 10 and 30 min for three different collision efficiencies. (From Warshaw, 1968.)

($E = 1$). The more accurate collision efficiencies of Davis and Sartor produce fewer large drops than the Hocking efficiencies. This arises from the fact that, though it has an 18-μm cutoff, the Hocking efficiency is larger in regions where $r/R \approx 0.5$, accounting for the slightly increased growth.

A convenient measure of the development of the large droplets in the spectrum is provided by the radius of the 100th largest droplet per cubic meter. At any time, 100 droplets per m³ are equal to or larger than this droplet. Figure 7.11 illustrates the growth of the 100th drop for different collision efficiencies and for two assumed initial droplet spectra. The spectrum of 50 droplets cm⁻³ approximates that of maritime cloud; that of 200 droplets cm⁻³ approximates continental cloud. Both have dispersions $\sigma/\bar{r} = 0.15$; both assume a liquid water content of 1 g/m³. The very slow evolution of continental as compared to maritime cloud is evident here.

These results of Warshaw, as well as the earlier findings of Telford and Twomey, are based in effect on the assumption of a

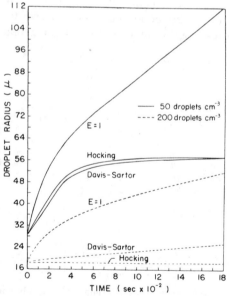

FIG. 7.11. Growth of the 100th largest droplet. (From Warshaw, 1968.)

well-mixed and infinite cloud. This is because no provision is made for the drops to fall out of the system by gravitational settling. Kovetz and Olund (1969) formulated more general equations for droplet growth which allow for settling in a finite cloud and account for growth by condensation as well as by coalescence. An example is shown in Fig. 7.12 of the development of the 100th largest drop in height and time in a cloud extending initially from 500 to 1000 m with an initial Gaussian distribution of droplets, the same as the maritime case considered by Warshaw. A uniform updraft of 10 cm/sec was assumed. By comparing cases with and without condensation included and for different collision efficiencies, Kovetz and Olund concluded that it was just as important to include condensation as to use a realistic collision efficiency. This conclusion is evident on comparing Figs. 7.11 and 7.12. After 600 sec, the 100th droplet of

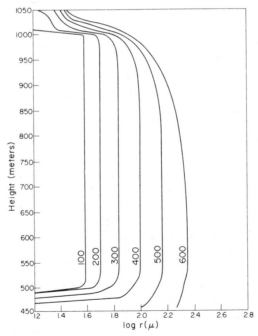

FIG. 7.12. Size of the 100th largest droplet at times indicated (sec). (From Kovetz and Olund, 1969.)

Kovetz and Olund has grown to a radius of about 200 μm (log $r = 2.3$), while that of Warshaw's maritime cloud has grown only to about 52 μm. Although the collision efficiencies were not the same for the two calculations this could not explain the wide disparity of results. Indeed, even for geometric sweepout ($E = 1$) Warshaw's 100th drop at 600 sec is only 72 μm in radius. Presumably, therefore, the faster growth in the Kovetz and Olund model is accounted for by condensation. However, it is possible that some of the difference in growth rate is due to different numerical integration and interpolation techniques.

Condensation plus stochastic coalescence

Anticipating the conclusion of Kovetz and Olund, East (1957) made calculations of droplet development by condensation and coalescence in moderate updrafts of several meters per second, finding that precipitation-size drops could be produced in realistically short periods of time. The stochastic coalescence equation was not actually used, the droplet population from the start being separated into collector drops and collected droplets. Also the collision efficiencies were inaccurately known. East's main contribution was to explain how condensation, which causes the droplet spectrum to narrow, actually speeds up the coalescence process. The spectrum narrows by condensation because small droplets grow (in radius) faster than large droplets. As the small droplets grow, their collision efficiency relative to the large drops increases at a rapid rate (see Fig. 7.4), and coalescence is thus accelerated. The relative velocity between large and small droplets, on which coalescence also depends, remains essentially unchanged in this narrowing process. Considering any two droplets of different initial size, we see from the approximate equation (6.19) that the difference between the squares of the radii remains constant. For droplet sizes less than about 40 μm the fall speed is given by Stokes' Law and the relative velocity between the droplets is constant. The net effect of condensation-narrowing is therefore to accelerate coalescence.

More exact calculations of droplet evolution by condensation and coalescence have been carried out by Leighton and Rogers (1974). Condensation was modeled by the approximation (6.19), and

Reinhardt's (1972) numerical techniques were used in solving the stochastic coalescence equation. This work was done to analyze droplet development in large convective clouds; consequently the main interest was in strong updrafts. Most calculations were made for 7.5 and 15 m/sec vertical air velocities. Accordingly, the condensation effects were more important than in previous work. Fallout of the larger drops was neglected, which is a better approximation for

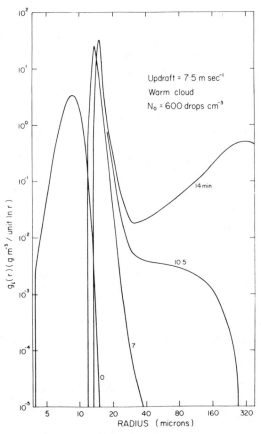

FIG. 7.13. Mass distribution after 0, 7, 10.5 and 14 min, corresponding to heights above cloud base of 0.7, 3.8, 5.3 and 7.0 km, for an updraft of 7.5 m/sec. (From Leighton and Rogers, 1974.)

strong updrafts than for weak ones. As another approximation, it was assumed that condensation occurs at a rate determined by the requirement that the amount of condensed water always equal the adiabatic liquid water content. This requirement then fixes the supersaturation.

Figure 7.13 gives an example of the results, in the form of droplet mass distributions at various times after a 7.5 m/sec updraft begins. The mass distribution $g_1(r)$ is such that $g_1(r)d(\ln r)$ equals the mass of cloud droplets per unit volume of air with radii in the interval $d(\ln r)$. (This distribution emphasizes the larger droplets much more than the spectrum function $n(r)$.) At 7 min condensation has shifted the distribution outwards on the radius axis and caused the formation of a pronounced peak at about 20 μm. Up to this time coalescence is negligible. After 7 min the distribution is sufficiently modified for coalescence to proceed, and the distribution at 10.5 min shows the development of a tail due to coalescence which extends to drizzle-drop size. From then on coalescence accelerates, but even at 14 min there is still a sharp peak arising from condensation which predominated earlier. Calculations for the same initial distribution in a 15 m/sec updraft show much less development at the 7 km level, since less time is available for coalescence growth.

As a test, calculations were made for the same initial distribution assuming coalescence only. After 14 min there was essentially negligible growth, confirming the importance of condensation.

Calculations similar to those of Leighton and Rogers, but with fewer approximations, have been reported by Young (1974). Starting with an activity spectrum of condensation nuclei, Young calculated the development of the droplet spectrum in an ascending element of air by condensation and coalescence, specifically accounting for the supersaturation. Figure 7.14 gives an example of his results, for an initial temperature and condensation nucleus population corresponding to a maritime cloud.

For the first 15 min of ascent there is no growth by coalescence: the droplet concentration remains unchanged as condensation is setting the stage for coalescence. Once it begins coalescence proceeds rapidly, as indicated by the abrupt decrease in the number of drops. At the same time the supersaturation increases sharply because the reduced number of drops are no longer able to

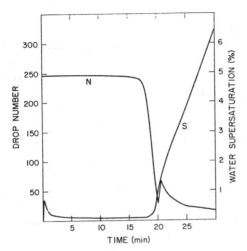

FIG. 7.14. The change with time of droplet concentration N and supersaturation S in a cloudy parcel ascending at 3 m/sec with an initial temperature of 15°C and a maritime-type spectrum of condensation nuclei. The droplet concentration is given for a mass of air which has a volume of 1 cm³ at the initial time and is therefore proportional to the droplet mixing ratio. (From Young, 1974.)

accommodate all the vapor made available by the updraft. The increasing supersaturation activates new condensation nuclei, at first causing a rise in drop number. This is only a transient effect, however, because the newly-formed droplets are quickly consumed by coalescence.

Young's calculations also take into account the breakup of drops once a large enough size is reached. This effect becomes important in establishing the distribution of raindrops with size, and is described in Chapter 9.

Concluding remarks

Observations show that rain can develop in warm clouds of the cumulus type in times as short as about 15 min after the cloud begins to form. It is generally agreed that the process responsible for this development must be gravitational coalescence among the droplets, in which case the droplet populations most likely to produce rain in a short time are those with broad enough spectra to have a high rate of collisions. A serious impediment to coalescence growth is the fact that collection efficiencies between small droplets are extremely small. Moreover, the condensation-diffusion process, which dominates droplet growth initially, leads to a narrowing of the spectrum, seemingly complicating the coalescence problem. The theoretical task is therefore to explain raindrop development in reasonable times in face of the small collection efficiencies and the approximate parabolic form of the diffusional growth law.

It is now recognized that statistical effects are crucial in the early stages of coalescence, so that the stochastic equation should be used to describe this process. Even so, coalescence alone is not sufficient to account for rain development in reasonably short times, starting with realistic droplet spectra. A number of mechanisms have been postulated for broadening the spectrum, thus paving the way for rapid coalescence. There is no agreement on which of these is the most important, although each may be effective under certain conditions. Rather than spectral broadening, it has been shown that the additional effect which can account for rain development, at least in some conditions, is continued condensation growth. This seems strange at first since condensation-diffusion, even with the kinetic corrections, leads to a narrowing of the droplet spectrum. Though the spectrum narrows the rate of coalescences increases because the collision efficiency increases rapidly as the droplets in the spectrum grow. Definitive calculations have not yet been made which account for rain development under various conditions of updraft speed, temperature, initial droplet spectrum, and entrainment mixing. Results from simple models are encouraging however, indicating that current theory may be on the verge of explaining the observations.

Problems

1. The liquid water content of a cloud 2 km in depth varies linearly from 1 g/m³ at the base to 3 g/m³ at the top. A drop of 100 microns diameter starts to fall from the top of the cloud. What will be its size when it leaves the cloud base? Assume that the collection efficiency is 0.8 and that there is no vertical air velocity. Neglect the fall speed and size of the cloud droplets.

2. A drop of 0.2 mm diameter is inserted in the base of a cumulus cloud that has a uniform liquid water content of 1.5 g/m³ and a constant updraft of 4 m/sec. Using the elementary form of the continuous-growth equation and neglecting growth by condensation, determine the following:

(a) The size of the drop at the top of its trajectory.
(b) The size of the drop as it leaves the cloud.
(c) The time the drop resides in the cloud.

For dependence of fall velocity on size use the data of Gunn and Kinzer (Table 7.2). Assume a collection efficiency of unity.

3. On a particular day the orographic cloud at the island of Hawaii is 2 km thick with a uniform liquid water content of 0.5 g/m³. A drop of 0.1 mm radius at cloud top begins to fall through the cloud.
(a) Find the size of the drop as it emerges from cloud base, neglecting vertical air motions in the cloud. In this and subsequent parts of Problem 3 neglect growth by condensation, use the elementary form of the continuous-growth equation, and assume a collection of efficiency of unity.
(b) Assuming that the terminal fall speed of the drop is equal to kr, where $k = 8 \times 10^3$ sec^{-1}, find the time required for the drop to fall through the cloud.
(c) Hawaiian orographic clouds are actually maintained by gentle upslope motions which cause a steady, weak updraft. Suppose that on the day in question there is a uniform updraft of 20 cm/sec. Find the size of the drop in part (a) as it emerges from the cloud, taking the updraft into account.

4. A drizzle-sized water drop is swept upwards in a cumulus congestus cloud and grows by accretion and condensation in the supersaturated environment. The supersaturation may be regarded as constant and the linear fall speed law of Problem 3 applies to the drop. Develop and solve the differential equation that describes the growth of this drop by accretion and condensation simultaneously. Compare the result with the approximation obtained by adding together the separate solutions for growth by accretion and condensation.

5. Calculate the distance through which a water drop, initially 0.1 mm in radius, falls before evaporating completely in an environment of 90% relative humidity and 278°K temperature. Neglect ventilation effects and use the Gunn and Kinzer data on fallspeed given in Table 7.2.

CHAPTER 8

FORMATION AND GROWTH OF ICE CRYSTALS

Nucleation of the ice phase

Once a cloud extends to altitudes above the 0°C level there is a chance that ice crystals will form. Two phase transitions can lead to ice formation: the freezing of a liquid droplet or the direct deposition (sublimation) of vapor to the solid phase. Both are nucleation processes, and in principle homogeneous and heterogeneous nucleation are possible.

In the same manner as for the nucleation of liquid, there are two equations that describe the homogeneous nucleation of ice, one for the size of the stable embryo, the other for the probability of occurrence of embryos due to the chance rearrangement of molecules. Both equations have a dependence on the surface free energy of a crystal/liquid interface, which is analogous to the surface tension in the equations for liquid nucleation. Until recently the theory of ice nucleation was incomplete because the value of the interfacial free energy was not accurately known. This introduces an uncertainty in calculations of the critical conditions for freezing nucleation. Analyzing some of the data on freezing temperatures of pure water droplets, Kuhns and Mason (1968) inferred that the value of the ice/liquid free energy must be approximately 20 erg/cm². Inserted in the equations, this value predicts that droplets smaller than 5 μm will freeze spontaneously at a temperature of about −40°C, in accord with most observations. Larger droplets are predicted to freeze at slightly warmer temperatures, also in agreement with observations. Hobbs and Ketcham (1969) have since measured the surface energy, determining a value of 33 ± 3 erg/cm² at 0°C. This value is consistent with that inferred by Kuhns and

116

Mason, as the free energy is expected to increase weakly with temperature.

Theory predicts that homogeneous nucleation by sublimation should occur only for extreme conditions of supersaturation. More than twenty-fold supersaturation with respect to ice is required at a temperature a few degrees below 0°C, and the critical value of supersaturation increases with decreasing temperature. According to Fletcher (1962), experiments on homogeneous sublimation indicate that the nucleation threshold is about −62°C with an eight-fold supersaturation with respect to water. Theory and experiments on this phase transition are complicated by the fact that under such extreme conditions homogeneous condensation, followed by freezing, is likely to occur and be confused with homogeneous sublimation. In spite of this uncertainty, it is clear that homogeneous sublimation cannot occur in the atmosphere, since such extreme supersaturations never exist.

On the other hand, natural clouds have been observed to contain some liquid droplets down to temperatures approaching −40°C, so it is likely that homogeneous freezing is a process that does occur in the atmosphere, at least in some clouds. The fact that natural clouds have a resistance to freezing is contrary to common experience, which indicates that water freezes when the temperature drops below 0°C. Our experience is based on observations of bulk water, in which a single nucleation event anywhere suffices to cause the entire mass to freeze. A cloud, however, is an unusual system in which the water mass is distributed over a large number of very small droplets, each one of which must experience a nucleation event somewhere within its boundaries before the cloud can become entirely frozen.

Ice crystals usually begin to appear in a cloud when the temperature drops below about −15°C. This implies that heterogeneous nucleation is occurring. Water in contact with most materials will freeze at temperatures warmer than −40°C, and sublimation will occur on most materials at supersaturations and supercoolings less than the homogeneous nucleation values. Thus the nucleation of ice in supercooled water or a supersaturated environment is aided by the presence of foreign surfaces or suspended particles.

Nucleation occurs most readily on surfaces having a lattice structure geometrically similar to that of ice. The material that most closely approximates ice in lattice structure, so far as is known, is silver iodide (AgI). In theory and experiments on heterogeneous ice nucleation there is again a problem in determining whether sublimation or condensation-freezing will occur in a given situation. A distinction between these processes is evidently possible for insoluble particles in the special case of an environmental vapor pressure above ice saturation but below water saturation. For this situation, at cold enough temperatures, nucleation may occur; it would be by direct sublimation since the condensation process (for an insoluble particle) would not be possible below water saturation. Generally speaking, condensation is more likely at small supercoolings and large supersaturation, while sublimation is more likely at large supercoolings and small supersaturations. Since the confusion exists, one often speaks of "ice nucleation" as the phenomenon, instead of the more specific "freezing nucleation" or "sublimation nucleation". Likewise, the atmospheric particles serving as nucleation centers can most safely be referred to as "ice nuclei".

Experiments on heterogeneous ice nucleation

Investigations of the growth of single crystals confirm the lattice-matching principle. Growth is found to occur most readily on nuclei which have a hexagonal lattice structure resembling that of natural ice, and slight solubility.

The nucleating properties of small particles are studied by introducing the particles into cloud chambers with controlled supercooling and supersaturation. The conditions are noted where the onset of nucleation occurs. (Ice crystals are usually discernable even in the presence of water cloud by the scintillation of the scattering from a strong beam of light.) In these experiments it may not be possible to distinguish between a sublimation event and a condensation event followed by freezing, except perhaps under the special conditions mentioned above. Table 8.1 indicates average results obtained by introducing various kinds of particles into a supercooled water cloud.

TABLE 8.1. *Nucleating Properties of Particles of Various Material*

	Material	Nucleation threshold (°C)
Inorganic	AgI	-4
	PbI_2	-6
	CuS	-7
Organic*	certain phenols	-2
		or warmer (Parungo and Lodge, 1965)
	phloroglucinol	-2
		(Langer *et al.*, 1963)

* For other comments on the effectiveness of organic materials see Fukuta (1966) and Braham (1963).

Another kind of experimental study of ice nuclei involves the freezing of water drops on a cold stage. To investigate natural atmospheric ice nuclei, rainwater (or hail or snow melt) is collected and then divided into a large number of small drops. The temperature is progressively lowered and the fraction of drops frozen as a function of temperature is noted. This gives an indication of the activity spectrum of freezing nuclei in the precipitation, from which may be inferred the concentration of freezing nuclei in the atmosphere (Vali, 1968). The same experiment may be used to assess the freezing-nucleation properties of any kind of particulate material, by mixing it with water, making the droplets, and cooling down to observe threshold freezing temperatures.

It should be emphasized that the droplet freezing experiments indicate only the *freezing* nuclei. Atmospheric ice nuclei are thought to be of three types, sublimation, freezing, and contact nuclei. A contact nucleus is one that causes ice to form by colliding with, but not necessarily becoming contained within, a supercooled droplet. The relative importance of these different kinds of nuclei is not yet clear.

Atmospheric ice nuclei

Several methods have been used to study atmospheric ice nuclei, the most common of which are cloud chambers and filter systems, into which samples of air are drawn. In the cloud chambers the sample is cooled down to a controlled temperature and a cloud is formed by adding sufficient water vapor. An optical system or some other means of ice crystal detection is used to count the number of crystals that form as a function of the degree of supercooling. This kind of experiment is not specific as to the type of nucleus (freezing, contact, or sublimation) and gives no information about nucleus size unless the incoming air sample is filtered to remove particles larger than a preselected size.

The second method involves the collection of aerosols by drawing the air sample through filter paper with known pore sizes. The particulates thereby trapped on the filters are then introduced to an environment suitable for ice crystal development and observations are made of the number of crystals that form. This technique does give information about the size of the nuclei but not about their mode of activation.

From these methods it has been determined that the concentration of ice nuclei is highly variable in space and time, and that a typical figure is 1 nucleus per liter at a temperature of $-20°C$. The concentration usually depends strongly on temperature, and a decrease of about $4°C$ causes the concentration to increase by an order of magnitude. This average temperature dependence is shown in Fig. 8.1. It must be realized that there can be wide deviations from the straight line that is shown.

Comparisons of ice nucleus counts with the concentration of crystals observed in clouds often indicate a wide discrepancy, the crystal concentrations exceeding those of the nuclei by two orders of magnitude or more. A possible explanation for the discrepancy is the ice crystal multiplication process (Hallett and Mossop, 1974), by which ice crystals colliding with and collecting relatively large cloud droplets at temperatures near $-5°C$ produce large numbers of secondary ice crystals. Other explanations have also been offered, as reviewed by Cotton et al. (1975). There is not yet a complete explanation of this disparity and it remains a fundamental unresolved problem.

Of primary interest are the nuclei which promote ice formation at relatively warm temperatures, for these are the ones that will initiate the first ice crystals in a developing cloud. These are generally believed to be within an order of magnitude of 1 μm in size, and to consist of kaolin (clay mineral) more often than not. These ideas are supported by the findings of Kumai (1967), who examined the

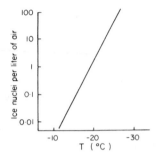

FIG. 8.1. Typical dependence of ice nucleus concentration on temperature.

particles found in snow crystals collected at the surface that appear
to be the growth centers of the crystals. For every snow crystal he
examined Kumai invariably found such a particle. Their sizes ranged
from 0.1 to 4 μm diameter and electron microscopy revealed them
usually to be silicates of aluminum (kaolin). This is a common
material found in many different soil types.

Vali's study of the freezing characteristics of rain led to different
conclusions. He found that a considerable number of the nuclei
were smaller than 0.01 μm in diameter and argued that other
methods of measuring nuclei systematically discriminate against
these small ones. Vali also deduced higher concentrations of nuclei
from his measurements than had earlier been reported. Precipitation
samples frequently contained numerous freezing nuclei active at
temperatures between -5 and $-10°C$. It was estimated that the
concentration of nuclei in the air in which the precipitation formed
may have been as high as 1 per liter for nuclei active at temperatures
warmer than $-10°C$ and 0.1 per liter for nuclei active above $-6°C$.
These estimates exceed the average conditions sketched in Fig. 8.1
by more than two orders of magnitude. Carrying out experiments on
suspensions of common soil particles, Vali found that freezing
nuclei active at temperatures as warm as $-6°C$ were not uncommon.
Separate experiments on suspensions of clay particles showed that
they were not so effective at the warm temperatures and that the
proportion of nuclei smaller than 0.01 μm was small. He concluded
that the atmospheric nuclei active at temperatures warmer than
$-10°C$ were smaller than 0.01 μm in diameter, not of clay origin,

and possibly associated with some minor organic component of soil.

There are uncertainties in Vali's results because only freezing nuclei are measured, and the atmospheric concentration must be inferred from precipitation samples by a rather indirect argument. Moreover, drop-freezing experiments only give the distribution of freezing temperatures for the particles at the time they are on the freezing stage. It is possible that the freezing properties of the particles could be modified as a result of immersion in the water. Nevertheless some of Vali's criticisms of the other techniques for measuring nuclei are valid, and it appears that much remains to be learned of atmospheric ice nuclei. From the meteorological point of view perhaps their most important characteristic is their relative scarcity. If we compare Vali's estimates of nucleus concentration (e.g. 1 per liter at $-10°C$) with the total number of Aitken particles in a typical sample (cf. Table 5.3) we see that only about one particle out of 10^6-10^9 is a freezing nucleus. (We should bear in mind also that Vali's estimates are higher than the others.) Looking at this a different way, while there is only one or so freezing nucleus per liter active at $-10°C$, there are in the order of 10^5 condensation nuclei per liter that are active for relative humidities slightly in excess of 100%. The atmosphere is thus characterized by a relative abundance of condensation nuclei and a scarcity of freezing nuclei. While supersaturations exceeding about 1% are extremely rare, supercooling to $-15°C$ or colder is not uncommon.

This section would be incomplete without a reference to the Bowen Meteor Hypothesis, which postulates an extraterrestrial source of ice nuclei. A discussion of the hypothesis is given by Fletcher (1962, pp. 250–258). Analyzing world-wide records of rainfall, Bowen found an apparent correlation between extreme rainfall occurrences and meteor showers. Both the statistical and the physical arguments in his hypothesis have been criticized, and generally the Meteor Hypothesis does not enjoy much support. The arguments were further weakened when Mason and others experimented with meteor dust in cloud chambers and found that it was not an especially effective nucleant. More recent work, however, indicates that the meteor question may not yet be laid to rest. Bigg and Guitronich (1967) vaporized meteorites at low pressure in a solar furnace, attempting to duplicate the small particles formed when meteors enter the atmosphere. The particles thus produced were found to be rather effective nucleants. Stony meteorites, for example, recondensed to form particles with dimensions of about 0.1 μm, producing 10^8–10^9 ice nuclei per gram active at $-10°C$. It was concluded that sufficient nuclei active at $-15°C$ are produced by meteors to account for the observed concentrations in the troposphere. This finding does not prove Bowen's hypothesis, but does indicate that at least some of the atmospheric ice nuclei may be of extraterrestrial origin.

Diffusional growth of ice crystals

Once ice embryos are formed, either by sublimation directly from the vapor or by the freezing of a supercooled droplet, diffusional growth will begin because the embryo is in an environment essentially saturated with respect to water. The growth equations are analogous to those in Chapter 6 for water droplets, but with an important difference due to the fact that the ice crystal is generally not spherical.

The diffusional growth problem is similar to problems in electrostatics and potential theory. By using Poisson's equation in electrostatics and Green's theorem, it can be shown that the integral over the surface of a conductor of the normal component of $-\nabla\phi$, where ϕ is the electrostatic potential, is equal to $4\pi C\phi_s$ where C is the capacitance of the conductor and ϕ_s its potential. If we identify $-D\nabla n$ which is the flux of water molecules with $-\nabla\phi$, then it follows that the total current of water out of the ice crystal is by analogy $4\pi CD(n_s - n_0)$ where n_s is the vapor number density at the crystal's surface and n_0 is its value far from the surface. The assumptions that $\nabla^2 n = 0$ and that n_s is the same over all points on the surface are necessary to complete the analogy. The generalized growth equation therefore becomes

$$\frac{dM}{dt} = 4\pi CD(\rho_v - \rho_{vr}). \qquad (8.1)$$

C denotes the electrical capacitance, with length units, a function of the size and shape of the particle. For a sphere, $C = r$ and (8.1) reduces to the growth equation for a water droplet. For a circular disk of radius r, which can be used as an approximation for plate-type ice crystals, $C = 2r/\pi$. Ice needles may be approximated by the formula for a prolate spheroid of major and minor semi-axes a and b, for which

$$C = \frac{A}{\ln[(a+A)/b]},$$

where $A = \sqrt{a^2 - b^2}$. For an oblate spheroid,

$$C = ae/\arcsin e,$$

where the ellipticity $e = \sqrt{1 - b^2/a^2}$.

Actual ice crystals have more complex shapes than the sphere, disk, and ellipsoids for which these theoretical formulas apply. Plane dendrites and plates, however, which are common crystal types, can be reasonably approximated by a circular disk of equal area. Likewise needles can be approximated by long prolates. According to Houghton (1950), who argued from the electrostatic analogy, the field about a plate covered with points is the same as that produced by a smooth plate a suitable distance away. The field is distorted by the irregularities only at distances from the plate which are the same order as the length and spacing of the ir-regularities. As the water vapor approaches the crystal, the fine structure determines how and where it will be deposited; but this does not affect the total flux of water vapor. Houghton's ideas were confirmed by laboratory measurements (McDonald, 1963b) of the capacitance of brass models of snowflakes.

As the ice crystal grows its surface is heated by the latent heat of sublimation and the value of ρ_{vr} is effectively raised above the value that would apply without heating. Under stationary growth conditions the value of ρ_{vr} is determined by the balance between the rates of latent heating and heat transfer away from the surface, which balance is expressed

$$\frac{\rho_v - \rho_{vr}}{T_r - T} = \frac{K}{L_s D}. \tag{8.2}$$

Following the thermodynamic arguments in Chapter 6, (8.1) and

(8.2) may be combined to give an analytical expression for crystal growth rate. This formula is precisely the same as was obtained for water drops if we formally interchange C for r, interpret $e_s(T)$ as the saturation vapor pressure over ice, and take L as the latent heat of sublimation:

$$\frac{dM}{dt} = \frac{4\pi C(S_i - 1)}{\left[\dfrac{L_s^2}{KR_vT^2} + \dfrac{R_vT}{e_i(T)D}\right]}. \tag{8.3}$$

Here, as in (6.17), the effects of surface free energy and ventilation are neglected. These effects are not understood as well for ice crystals as for water droplets. Vapor molecules cannot unite with an ice crystal in any haphazard way, but must join up, molecule-by-molecule, in such a manner that the crystal morphology is maintained. Consequently it may be incorrect to identify ρ_{vr} with the equilibrium vapor density of ice; and in fact ρ_{vr} may not be the same over all points of the crystal surface. Because of these effects the rate of growth of an ice crystal will tend to be slower than given by (8.3). Experiments by Fukuta (1969) indicated that at temperatures between about $0°$ and $-10°C$ the growth rates of small crystals are about half as fast as predicted by (8.3). For large crystal sizes, the formula may be a better approximation.

In (8.3) S_i refers to the ambient saturation ratio with respect to ice, which may be written

$$S_i = S(e_s/e_i) \tag{8.4}$$

where S is the ambient saturation ratio with respect to water, and e_s and e_i are the equilibrium vapor pressures over water and ice, respectively. As may be confirmed from Table 2.1 of Chapter 2, the ratio e_s/e_i exceeds unity for all (subfreezing) temperatures, and increases approximately linearly with decreasing temperature to a value of about 1.5 at $-40°C$. Since the initial growth of ice crystals usually occurs in a water cloud, $S \approx 1$ in (8.4) and $S_i \approx (e_s/e_i)$. Consequently the temperature dependence of (e_s/e_i) exerts a strong influence on crystal growth rate dM/dt. The bracketed term in the denominator of (8.3) depends on temperature and pressure in approximately the manner indicated in Fig. 6.1 for the case of water droplets. Byers (1965), has combined this temperature dependence with that of S_i, assuming a water-saturated environment, to determine the dependence of ice crystal growth rate on temperature and pressure as shown in Fig. 8.2. These curves indicate that the growth rate varies inversely with pressure, and that the temperature for maximum growth is about −15°C over a wide range of pressure.

Ambient conditions determine not only growth rate, but also the form, or habit, that a growing crystal takes. Figure 8.3 illustrates the main crystal types, namely column, plate, and dendrite. As a growing crystal descends through cloud its crystal habit will change

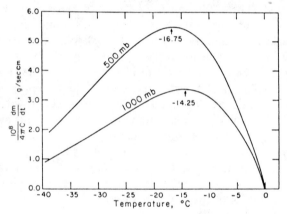

FIG. 8.2. Normalized ice crystal growth rate as a function of temperature. (Adapted from Byers, 1965.)

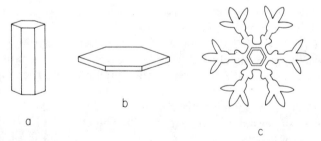

FIG. 8.3. Schematic representation of the main shapes of ice crystals: (a) column; (b) plate; (c) dendrite.

according to the changing ambient conditions. Sector stars are formed when plates develop peripheral dendritic structure; capped columns arise when columns develop plates on their ends. The intricate stellar shapes which are often observed are variations on the dendritic form.

Since the pioneering work of Nakaya (1954) several authors have reported on the dependence of crystal habit on environmental conditions. Their work is in good agreement, and is summarized in Fig. 8.4, due to Kobayashi (from Fletcher, 1962, p. 265). This figure also indicates the excess vapor density over ice equilibrium in an atmosphere saturated with respect to plane water. This approximates the conditions to be expected in clouds, and shows that the excess vapor density is a maximum at about $-15°C$, which is also the temperature of maximum growth rate. The preferred crystal types in this favored growth region are seen to be dendrites and sectors.

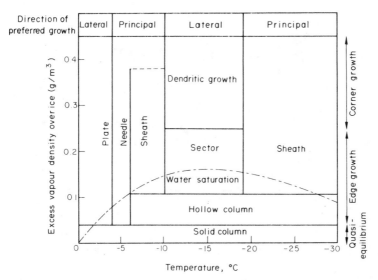

FIG. 8.4. Kobayashi's diagram of crystal habit as function of temperature and excess vapor density over ice saturation. (From Fletcher, 1962.)

Further growth by accretion

An ice crystal falling through a cloud of supercooled water droplets and other ice crystals may grow by the accretion of water or by aggregation with other crystals. Accretional growth leads to rimed structures and graupel; aggregation leads to snowflakes. Of importance in growth by sweepout is the fall speed of ice crystals, some of the classical data on which are given in Fig. 8.5, from Fletcher (1962, p. 276). The fastest falling crystals are seen to be graupel particles (which are not really crystals, but aggregates of frozen droplets). The rimed structures (crystal with droplets) fall at about 1 m/sec, but all the pure crystal types fall at less than 1 m/sec. An empirical formula that provides an approximate fit to the graupel curve is

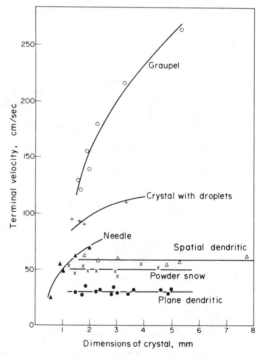

FIG. 8.5. Nakaya and Terada's measured terminal velocities of ice crystals. (From Fletcher, 1962.)

$$u = 520r^{0.6} \tag{8.5}$$

with u in cm/sec and r, the radius of the sphere which just circumscribes the particle, in cm. Snowflakes are not shown on this figure but they are also found to fall at about 1 m/sec so long as they are dry. When melting begins they become more compact and fall faster. A generally accepted approximation for snowflake fall speed is due to Langleben (1954),

$$u = kD^n, \tag{8.6}$$

where D is the melted diameter. With D in cm and u the fall speed in cm/sec, Langleben found for dendrites $k \approx 160$ and $n \approx 0.3$, and for columns and plates $k \approx 234$ and $n \approx 0.3$.

Mason (1971) gave empirical formulas relating the mass of ice crystals to the diameter of their circumscribing spheres, of the form

$$m = aD^b. \tag{8.7}$$

Table 8.2 gives the values of a and b for various crystal types. There are such wide variations in habit and density of ice crystals that (8.7) must be viewed only as a rough approximation.

TABLE 8.2. *Values of a and b in (8.7), for D in cm and m in g*

Crystal type	a	b
Graupel	0.065	3
Rimed plates and stellar dendrites	0.027	2
Powder snow and spatial dendrites	0.010	2
Plane dendrites	0.0038	2

In the process of growth by accretion the question of collection efficiency arises. First there is the aerodynamic problem of collision efficiency; then the question of whether sticking occurs, given a collision. Not much is known about either side of the question. Since ice crystals fall more slowly than water droplets of equal mass, it seems plausible that the collision efficiencies might be higher. However, Pitter and Pruppacher (1974) have determined collision efficiencies of simple ice plates theoretically, by calculating trajectories of water droplets relative to the ice crystals, and have shown the collection process to be quite complex, with the perimeter of the crystal a preferred area for collisions. Since freezing is likely to occur on contact with supercooled droplets, the coalescence efficiency might be expected to be unity.

In the process of crystal aggregation, the collection efficiency is less well understood. Indications are that open structures like dendrites are more likely to stick, given a collision, than crystals of other shape, and that sticking in any case is more likely at relatively warm temperatures. Judging from the observed sizes of snowflakes as a function of temperature, it has been inferred that significant aggregation is possible only at temperatures warmer than −10°C.

Bearing the uncertainties in mind, one can set up equations to describe growth by these processes. For the case of accretional growth, leading to graupel, an approximation analogous to (7.15) may be employed,

$$\frac{dm}{dt} = \bar{E} M \pi R^2 u(R), \qquad (8.8)$$

where m = mass of particle, M = cloud liquid water content, R = radius of particle, $u(R)$ = fall speed, and $\bar{E}$ = mean collection efficiency.

Basically the same approach may be used in analyzing the aggregation process. Since snowflakes all fall at about 1 m/sec and ice crystals all fall at about 0.4–0.5 m/sec, the growth equation for a snowflake is

$$\frac{dm}{dt} = \bar{E} M \pi R^2 \Delta u. \qquad (8.9)$$

where Δu is the difference in fall speed of the snowflake and the ice crystals, essentially a constant. Sometimes the population of ice crystals is more conveniently characterized by the number density N than by the density of frozen water M. They are related by

$$M = N v \rho,$$

where v is the (average) volume of the crystals and ρ is their density. If the snowflake is assumed to have the same density, then $m = \rho V$ where V denotes its volume. The growth equation in terms of volume then becomes

$$\frac{dV}{dt} = B\bar{E}V^{2/3}Nv\,\Delta u, \tag{8.10}$$

where $B^3 = \frac{9}{16}\pi$. Clearly these equations must be understood as rough approximations to the actual growth processes. According to Fletcher (1962), calculations based on (8.10) have been found to give results in reasonable accord with observations on graupel and snowflakes.

Fundamentally, snowflakes must develop because a few of the crystals, which formed and grew by diffusion, become larger than their neighbors, either by enhanced diffusional growth or by chance collisions with other crystals or supercooled droplets. Once having attained this initial advantage, the crystals or small aggregates are in a favorable position to grow by the sweepout process. A complete theory for the development of precipitation in the ice phase should therefore take into account the statistical effects that are incorporated in the coalescence theory of rain (see Chapter 7). The only approach of this kind up to now was reported by Austin and Kraus (1968). They formulated a model for snowflake development on the assumption that random collisions lead to a distribution of aggregate sizes such that gravitational effects can become important. Taking initial values of number density between 10^4 and 10^5 crystals per cubic meter and assuming about 100 collisions per second per cubic meter, they found that reasonable distributions of snowflakes resulted in realistic time periods. The results depend rather critically on the collision frequency, and there seems to be no independent means of establishing 100 as a typical figure for this frequency.

The ice crystal process versus coalescence

In order for raindrops or snowflakes of appreciable size to develop, it is necessary that aggregation or accretion take place in the case of ice-phase growth, or that coalescence take place in the all-water process. Condensation-diffusion alone cannot explain the formation of 2 and 3 mm raindrops in the time available for growth. This process is more effective for ice crystals than for water droplets, however, owing to the fact that the cloudy environment tends to be just saturated with respect to water and supersaturated with respect to ice. It is common experience that light precipitation can occur in the form of individual ice crystals, indicating that aggregation or accretion never occurred. It is reasonable to suppose, therefore, that some of the precipitation reaching the surface in the form of drizzle or very light rain might be due to unaggregated crystals that melt before reaching the ground. In warm clouds on the other hand diffusional growth is too slow to produce even drizzle-size drops in reasonable times; coalescence is always required to produce rain from such clouds.

Many cumulus clouds develop initially at temperatures warmer than 0°C, or at least warm enough to make droplet freezing unlikely, and then grow vertically to altitudes considerably above the 0°C level, where ice crystal formation is likely to occur. In such clouds, both precipitation mechanisms may occur—initially the coalescence process among droplets, later the ice crystal process as well. Which process dominates in a given situation depends primarily on the temperature at cloud top, the cloud liquid water content, and to some extent the droplet concentration. The coalescence process will tend to predominate in clouds that are relatively warm with high liquid water contents and low droplet concentrations.

A comparison of the rates of precipitation formation in the two processes was made by Houghton (1950). Using the best available information on cloud droplet spectra and collection efficiencies, he calculated the growth by coalescence of the largest droplet present. Against this he compared the growth by diffusion of ice crystals of various shapes, assuming them to be in an environment saturated with respect to water. The results for coalescence showed the rate of growth dm/dt to be proportional to cloud water content M. For both calculations he found that

$$\frac{1}{m}\frac{dm}{dt} \propto m^k$$

to a good approximation, where k was larger for coalescence than for diffusion. Fig. 8.6, based on Houghton's results, illustrates $(1/m)(dm/dt)$ for an ice crystal growing by diffusion and a droplet growing by coalescence. Both are favorable cases, the water droplet coming from a broad distribution with $M = 1 \text{ g/m}^3$, and the ice crystal being a hexagonal plate growing at −15°C. This comparison shows that the ice crystal process outpaces coalescence at small radii, but that coalescence becomes more rapid for radii larger than

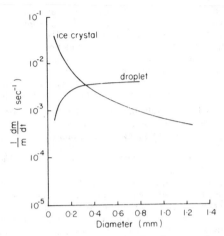

FIG. 8.6. Comparison of relative growth rates of an ice crystal and a water droplet, under favorable conditions for both. (From data in Houghton, 1950.)

about 0.3 mm—the size of large drizzle drops. Because of the slowness of initial coalescence growth, Houghton concluded that precipitation was most likely initiated by the ice crystal process in typical midlatitude cumulus clouds, but that a sweepout process, either aggregation of the crystals or accretion by crystals, snowflakes, or melted snowflakes, was required to produce large raindrops.

Radar observations since the time of Houghton's work have indicated that the first (radar detectable) precipitation often appears at cloud levels warmer than 0°C. This strongly suggests that the coalescence process initiates precipitation, although once ice crystals form there is no doubt that they accelerate the process.

Problems

1. A laboratory experiment with an ice crystal of hexagonal-plate form is carried out under carefully controlled conditions. At a pressure of 1000 mb and a temperature of −10°C, the ice crystal is observed to sublimate completely in 10^4 sec (≈ 2.8 hr),

from an initial radius of 5 mm. Assuming balanced and isothermal conditions throughout the process, compute the temperature of the ice crystal. Take the bulk density of ice as 0.8 g/cm^3 and assume the thickness of the plate is one-tenth its radius.

2. The classical equation for diffusional growth of a water droplet or ice crystal takes into account heat transfer from the particle to its environment by conduction, but neglects heat transfer by radiation. Consider a droplet growing by diffusion of vapor and compare the rate of heat loss by conduction with the rate of heat loss by radiation to the ambient air. Assume that the droplet and the ambient air may be approximated as black body emitters. From your result does it appear safe to neglect radiation in calculating the diffusional growth of a droplet? Is it equally safe to neglect radiation in the case of ice crystal growth by diffusion?

3. It has been reported (R. R. Braham, *J. Atmos. Sci.*, May 1967) that ice crystals from high-level cirrus clouds were observed to "trigger" the development of mid-level clouds after surviving a long descent through cloud-free air. In one case, numerous crystals survived a fall of 5 km in air with a relative humidity less than 20%. Is this observation consistent with the equation (8.3) describing diffusional growth (and sublimation) of ice crystals?

Approach the question the following way:

Assume the air to be isothermal at $-15°C$ with a constant relative humidity of 20% (with respect to water saturation). Assume that the ice crystals fall at a constant speed of 50 cm/sec independently of their size, that they are in the form of hexagonal plates, and that mass and radius are related (in c.g.s. units) by

$$m = 1.5 \times 10^{-3} r^2.$$

Determine the size that such crystals must have initially to survive the 5 km fall. Does this size seem reasonable for the ice crystals in a cirrus cloud? As an approximation, assume that the process takes place at $p = 600$ mb.

CHAPTER 9

RAIN AND SNOW

Drop-size distribution

Precipitation may be initiated through either the coalescence process or the ice-crystal process, with coalescence favored in clouds that are relatively warm with high liquid water contents. After precipitation particles are formed they grow primarily by sweeping out cloud droplets (accretion) or by combining with one another. Depending upon various factors, this continued growth produces raindrops, snowflakes, or hail.

Regardless of how it is initiated, precipitation over much of the world reaches the ground as rain. Its most commonly measured characteristic is the rainfall rate at the surface. A more complete description of the rain is provided by the drop-size distribution function, which expresses the number of drops per unit size interval (usually diameter) per unit volume of space. Such distributions have been measured by a variety of methods in most of the world's climatic regions. Though they are variable in time and space, the distributions usually indicate a rapid decrease in drop concentration with increasing size, at least for diameters exceeding about 1 mm. Also they generally show a systematic variation with rainfall intensity, the number of large drops tending to increase with rainfall rate.

Some examples of raindrop spectra are shown in Fig. 9.1. These were obtained with an instrument having a collection area of 50 cm^2 which records the size of individual raindrops by sensing their momentum on impact. For each curve the sample time is indicated as well as the total number of drops counted. A relatively large sample is needed when estimating drop-size distributions to suppress the statistical variability in counts of the rare large drops. In this figure the distributions numbered 1 and 2 were recorded in steady rain; distribution 3 was measured in a thunderstorm.

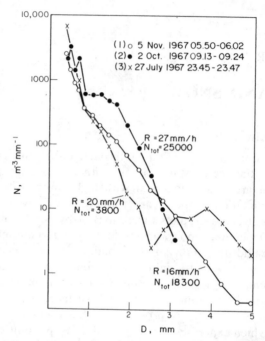

FIG. 9.1. Examples of measured drop-size distributions in rain. Indicated for each curve are the duration of the observation, the total number of drops counted, and the average rainfall rate. Distributions 1 and 2 were recorded during nearly constant rain; distribution 3 was recorded during a thunderstorm. (From Joss *et al.*, 1968.)

These examples, and the measurements of many others, indicate that drop-size distributions are of an approximate negative-exponential form, especially in rain that is fairly steady. Marshall and Palmer (1948) first suggested this approximation on the basis of a summer's observations in Ottawa, Canada. Figure 9.2 compares drop spectra at three values of rainfall rate with the best-fit exponential approximations, which are straight lines on the semilogarithmic coordinates. Thus the drop-size distributions, except for very small sizes, may be approximated as

$$N(D) = N_0 e^{-\Lambda D}, \tag{9.1}$$

where $N(D)dD$ is the number of drops per unit volume with

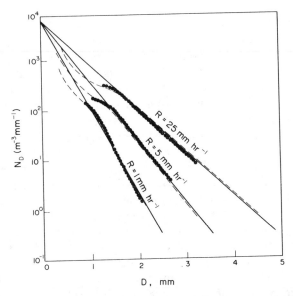

, FIG. 9.2. Measured drop-size distributions (dotted lines) compared with best-fit exponential curves (straight lines) and distributions reported by others (dashed lines). (From Marshall and Palmer, 1948.)

diameters between D and $D + dD$. Marshall and Palmer found that the slope factor Λ depends only on rainfall rate and is given by

$$\Lambda(R) = 41 R^{-0.21}, \tag{9.2}$$

where Λ has units of cm^{-1} and R is measured in mm/hr. Rather remarkably, they also found that the intercept parameter N_0 is a constant given by

$$N_0 = 0.08 \, cm^{-4}. \tag{9.3}$$

It is obvious from Fig. 9.1 that not all drop-size distributions have the simple exponential form. Yet measurements from many different regions have shown that an exponential tends to be the limiting form as individual samples are averaged. Moreover, for steady rain at continental midlatitudes, the Marshall–Palmer values $_0$ of Λ and N_0 are often found to be reasonable approximations.

Drop breakup

An explanation of the tendency for drop-size distributions to approach a negative-exponential form is provided, at least in part, by the phenomenon of breakup. Raindrops are limited in size because the chance of disruption increases with size. One cause of breakup is the aerodynamically induced circulation of water in the drop. Once a diameter of about 3 mm is attained it is no longer sure that surface tension can hold the drop together; a drop as large as 6 mm in diameter is unstable and can exist only briefly before breaking apart. Upon breakup a number of smaller drops are produced. Komabayasi *et al.* (1964) have given data on the probability of spontaneous breakup as a function of drop size, and the size spectrum of small drops thereby produced.

Another cause of breakup is collisions between drops. In a laboratory study of water drops of diameter 0.3 to 1.5 mm colliding with relative velocities ranging from 0.3 to 3 m/sec, Brazier-Smith *et al.* (1972) found that permanent coalescence becomes less likely for increasing values of drop size, relative velocity, and impact parameter. Collisions at grazing incidence produce a spinning, elongated drop which may quickly fly apart, resulting in the formation of satellite drops. Disruption occurs when the rotational kinetic energy of the coupled drops exceeds the surface energy required to produce separate drops. From a comparison of these energies, Brazier-Smith *et al.* obtained the following expression for the coalescence efficiency:

$$\varepsilon = \frac{12\sigma f(R/r)}{5r\rho U^2} \tag{9.4}$$

where σ is the surface tension of the drops, ρ the density, U the relative velocity, and $f(R/r)$ a dimensionless factor given by

$$f(R/r) = f(\gamma) = \frac{[1 + \gamma^2 - (1 + \gamma^3)^{2/3}][1 + \gamma^3]^{11/3}}{\gamma^6(1 + \gamma)^2}. \tag{9.5}$$

For the range of drop sizes tested good agreement was found between (9.4) and the laboratory observations. It should be noted that ε defined by this equation can take on values greater than unity for certain combinations of R, r, and U. Evidently ε may be interpreted as the coalescence efficiency only for values less than or equal to unity.

For drops freely falling in the atmosphere the relative velocity U is the difference in terminal fall speed of drops of radius R and r. The Gunn and Kinzer data of Table 7.2 were used to determine U for all radius pairs. Then (9.4) was evaluated as a function of R and r, and is plotted in Fig. 9.3. This figure shows that the coalescence efficiency is unity for $R < 0.4$ mm or $r < 0.2$ mm. Values of ε less than 0.2 occur for 1 mm $< R < 2.5$ mm. Minima of ε fall approximately on the line $r = 0.6R$. The efficiency is unity for all values of r

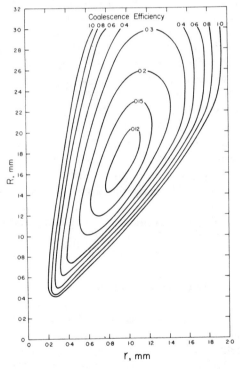

FIG. 9.3. Coalescence efficiency from the theory of Brazier-Smith *et al.*, as a function of drop radii. The value of ε is the fraction of collisions that result in permanent coalescence. All collisions lead to coalescence outside the contour $\varepsilon = 1$.

such that $r < 0.2R$ or $r > 0.9R$. Comparing Figs. 9.3 and 7.4 indicates that the collision efficiency is essentially unity in the region where the coalescence efficiency is less than unity, and conversely. Therefore the *collection* efficiency, to good approximation, equals either the collision efficiency or the coalescence efficiency, whichever is the smaller.

Brazier-Smith *et al.* observed that most of the collisions that did not lead to permanent coalescence resulted in the production of small satellite drops, ranging in number·from 1 to 10 and with sizes from 20 to 220 μm radius. In a theoretical study of raindrop interactions and rainfall rates within clouds, the same authors (Brazier-Smith *et al.*, 1973) explained that breakup may be approximated reasonably well by assuming that every collision-induced disruption produces three satellite drops of equal size, each with a volume given by $0.04 V_1 V_2/(V_1 + V_2)$ where V_1 and V_2 denote the volumes of the parent drops. Because of the limited range of radius pairs over which collision-breakup can occur, this formula usually predicts satellite drops of about 100 μm radius.

Srivastava (1971) attempted to determine theoretically whether raindrop coalescence and breakup would lead to a Marshall–Palmer distribution. He applied the stochastic coalescence equation to an assumed initial distribution of small raindrops and included spontaneous breakup according to the data of Komabayasi *et al.*, in slightly revised form. The drop-size distributions that resulted were not found to fit the simple negative-exponential law, being rather too flat in the size range from about 2 to 5 mm diameter. Strivastava suggested that the failure to approach exponential form might have arisen from the neglect of condensation or collision-induced breakup.

In a recent theoretical treatment of drop-spectrum evolution, Young (1975) has included the effects of collision-induced breakup along with growth by condensation and coalescence. Starting with an assumed activity spectrum of condensation nuclei, he calculated the drop-size distribution as a function of time in a volume of cloudy air ascending at constant velocity. The breakup process was modeled according to the findings of Brazier-Smith *et al.* Thus, after drops grow beyond 0.4 mm some fraction of the collisions cause fragmentation and the production of satellite drops. Figure 9.4 shows

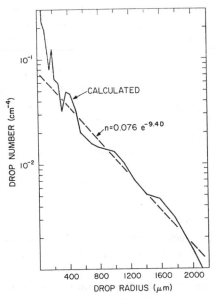

FIG. 9.4. Theoretical drop-size distribution produced after 30 min in a model of droplet growth that includes the effects of condensation, coalescence, and collision breakup. (Adapted from Young, 1975.)

the raindrop spectrum after 30 min of ascent at 3 m/sec in a cloud with maritime characteristics. The spectrum can be adequately approximated by an exponential of the form (9.1). Although Young pointed out that the exponent Λ in the best-fit curve differs from that predicted by (9.2) for the calculated rainfall rate, the results convincingly demonstrate the importance of collision breakup in establishing the limiting exponential form. The discrepancy in Λ should not be too disturbing in any case, because the Marshall–Palmer relation (9.2) has not been found to fit maritime rain.

Distribution of snowflakes with size

Snowflakes rather than individual ice crystals account for most of the precipitation reaching the ground as snow. As the snowflakes are irregular aggregates of crystals or smaller snowflakes, there is no easy way to measure their linear dimensions. Consequently, data on snowflake sizes are usually expressed in terms of particle mass or, equivalently, the diameter of the water drop formed when the snowflake melts.

Size distributions of aggregate snowflakes were measured by Gunn and Marshall (1958). They plotted the data on semilogarithmic coordinates, as for raindrops, obtaining the results shown in Fig. 9.5.

Once again the data points for a given rate of precipitation can be fitted reasonably well by an exponential function of the form (9.1). For snow, however, the parameters are related to precipitation intensity by

$$\Lambda(\text{cm}^{-1}) = 25.5 R^{-0.48} \tag{9.6}$$

and

$$N_0(\text{cm}^{-4}) = 3.8 \times 10^{-2} R^{-0.87}. \tag{9.7}$$

In these equations the precipitation rate R (mm/hr) is in terms of the water-equivalent depth of the accumulated snow.

In theoretical studies of precipitation development it is often necessary to compute moments of the drop-size distribution. For example, the flux of precipitation through a horizontal area, the

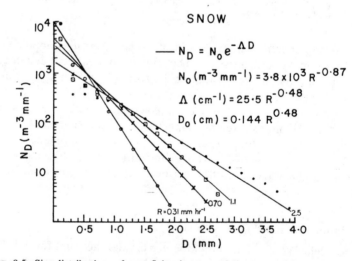

FIG. 9.5. Size distributions of snowflakes in terms of diameters of drops produced by melting the snowflakes. (From Gunn and Marshall, 1958.)

mass of precipitation per unit volume, and the radar reflectivity of the precipitation are all simply related to certain moments of $N(D)$. This makes the exponential approximation especially convenient to use in theoretical work, since the moments are known analytically: the nth moment is given by

$$\int_0^\infty D^n N(D)\,dD = N_0 \frac{\Gamma(n+1)}{\Lambda^{n+1}},\qquad (9.8)$$

where Γ denotes the gamma function. (For n an integer $\Gamma(n+1) = n!$.) This analytical result, which requires an infinite upper limit of integration, is usually a good approximation for real distributions that have a finite upper limit of diameter. The exponential form of $N(D)$ falls off so rapidly with D that the unrealistic large particles implied by the infinite limit make little contribution to the integral.

Reanalyzing the data of Gunn and Marshall and of others, Sekhon and Srivastava (1970) determined that the negative-exponential function (9.1) provides an adequate fit for all the observations, but that more consistent results can be achieved in theoretical work with the moments of $N(D)$ if the parameters have the values

$$\Lambda(\mathrm{cm}^{-1}) = 22.9 R^{-0.45} \qquad (9.9)$$

and

$$N_0(\mathrm{cm}^{-4}) = 2.5 \times 10^{-2} R^{-0.94}. \qquad (9.10)$$

Because of uncertainties in the methods of measuring $N(D)$ for snowflakes some discrepancies in results are expected. The measurements are obtained from observations of the number of flakes of a given size that fall on a horizontal surface during a certain exposure time. To infer from these observations the concentration $N(D)$ of snowflakes in space, it is necessary to divide the observed distributions by the fall speeds of the snowflakes in each size interval. These fall speeds are not uniquely dependent on size, but depend also upon density and possibly the crystal forms that make up the snowflake. Uncertainty is thus introduced into the estimates of $N(D)$, undoubtedly accounting for some of the disagreement in data given by different investigators. Moreover, since there is real variability in snowflake structure there is quite likely to be real variability in the dependence of $N(D)$ on precipitation rate. Without other information, however, it is probably reasonable in theoretical studies to use the formulas of Sekhon and Srivastava, which were found to fit a large number of observations adequately and to lead to consistent relationships among the moments of $N(D)$.

Aggregation and breakup of snowflakes

Jiusto and Weickmann (1973) found that the larger snowflakes—those consisting of 10 to 100 or more individual crystals—often consist of dendrites and thin plates indicative of diffusional growth conditions near water saturation. Column and thick-plate aggregates, signifying ice-saturation conditions, are far less common. Rimed crystal forms, which indicate growth by accretion of supercooled droplets, are common when ambient temperatures are relatively mild, and tend to be produced by convective clouds, suggesting the requirements of high liquid water contents and a scarcity of ice nuclei. Jiusto and Weickmann also reported that irregular crystal forms such as fragments, rimed branches, and nonsymmetric segments are sometimes dominant in heavy snowfall.

While it is obvious enough that snowflake size distributions, like those of raindrops, are established by the processes of growth and fracturing, the development of snowflake populations is more difficult to analyze theoretically. The crystal habit is significant in determining the diffusional growth rate, and may also affect the tendency for clumping. Fracturing of snowflakes is likely to be collision-induced, but it may depend on crystal type and temperature. General equations for ice crystal growth by accretion and aggregation have been formulated (see pp. 128–129), but with the many uncertainties about crystal formation and interaction it is unlikely that a comprehensive model of the evolution of size distributions in snow will be developed soon.

Precipitation rates

The precipitation rate or intensity is the flux of precipitation through a horizontal surface. It is measured in terms of the volume flux of water. The c.g.s. units are therefore $cm^3 cm^{-2} sec^{-1} = cm sec^{-1}$, but by convention it is usually expressed in mm/hr.

The intensity can be written in terms of the size distribution function $N(D)$ as

$$R = \frac{\pi}{6} \int_0^\infty N(D)D^3 u(D) dD, \qquad (9.11)$$

where $u(D)$ is the fall velocity of particles of size D. With the convention that D refers to melted diameter and R to the equivalent rainfall rate, (9.11) applies to snow as well as rain.

At levels above the ground it is possible that vertical air motions are present in which case the interpretation of precipitation intensity as a flux becomes ambiguous. With an updraft of velocity U present, the flux of precipitation becomes

$$R = \frac{\pi}{6} \int_0^\infty N(D)D^3 (u - U) dD. \qquad (9.12)$$

Obviously this quantity can become negative for U sufficiently large. A measure of the precipitation amount that is independent of the updraft speed is the precipitation water content L, defined by

$$L = \frac{\pi}{6} \rho_L \int_0^\infty N(D)D^3 dD. \qquad (9.13)$$

Values of R at the surface can vary from trace amounts up to several hundred mm/hr. Rainfall rates in excess of about 25 mm/hr are always associated with convective clouds. At most localities precipitation rates in the form of snowfall tend to be at least an order of magnitude less than those in the form of rainfall. For the Montreal area, which may be fairly typical of many midlatitude localities, the main contribution to a year's total rainfall comes from rainfall rates of about 10 mm/hr.

Analyzing rain development by condensation, coalescence, and breakup, Brazier-Smith *et al.* (1973) found that the cloud water content, rainwater content, and precipitation rate are all insensitive to the coalescence efficiency and to the production of satellite drops upon breakup. While microphysical processes are all-important in the development of the size distributions of precipitation particles, the intensity and duration are largely controlled by kinematic and thermodynamic factors, especially the temperature at cloud base, the cloud thickness, and the updraft speed.

Problems

1. A small graupel particle is caught in the updraft of a developing cumulonimbus cloud. Its initial diameter is 0.1 mm and the updraft is constant at 10 m/sec. Determine the time required for the particle to grow to a diameter of 2 mm and the distance through which it ascends during this time. Assume the cloud liquid water content is constant at 2 g/m^3. Neglect growth by sublimation.

   The relation between the mass and radius of graupel is approximately

$$m = 0.52r^3 \quad \text{(c.g.s. units)}$$

   and the dependence of graupel fall speed on size may be approximated as

$$u(r) = 520r^{0.6} \quad \text{(c.g.s. units)}.$$

   Assume a collection efficiency of unity.

2. As an index to the efficiency of precipitation growth by accretion, H. G. Houghton (*J. Appl. Meteor.*, October 1968) considered the total fraction of a unit horizontal area in a cloud that is geometrically swept by the precipitation particles in a unit time. Derive an expression for this efficiency as a function of rainfall rate for the following precipitation model:

   (i) Drop-size spectrum of the general negative-exponential form,

$$N(D) = N_0 \exp(-bD),$$

   with N_0 a constant equal to 0.08 cm^{-4} and b a parameter depending on rainfall rate;

   (ii) Terminal fall velocity related to drop size by

$$V(D) = kD,$$

   with $k = 4 \times 10^3 \text{ sec}^{-1}$;

   (iii) Zero updraft velocity.

   Evaluate your result for a rainfall rate of 20 mm/hr.

3. The rate of production of cloud water at a given level in a developing convective cloud is proportional to the updraft speed at that level. As rain falls through the level it sweeps out cloud droplets and depletes the cloud water content. Derive the relationship between rainfall rate and updraft speed that corresponds to a balance between the rate of cloud water production and the rate of sweepout. Make the following assumptions:

   (1) All vapor excess over the equilibrium saturation ratio immediately condenses to form cloud droplets.
   (2) Raindrop size distribution of the negative-exponential form as in Problem 2.
   (3) Terminal fall speed linearly related to raindrop diameter as in Problem 2.

CHAPTER 10

WEATHER RADAR

DURING World War II it was discovered that the newly developed microwave radars, designed for spotting distant ships and airplanes, were sometimes hampered in their operation by the presence of weather "clutter". Extensive theoretical and experimental work in the middle and late 1940's showed that the clutter arose from electromagnetic scattering by precipitation. The early findings have since been refined and elaborated to the point that most of the measurable properties of radar waves returned from precipitation can be interpreted in terms of the sizes, shapes, motions, or thermodynamic phase of the precipitation particles. Radars have become essential observational tools in studies of storm development and precipitation growth, and entire radar systems complete with computer data processors have been built for meteorological research.

This chapter outlines the essentials of radar meteorology and provides a background for the radar illustrations in subsequent chapters. A comprehensive treatment of radar principles and meteorological applications is given by Battan (1973).

Principles of radar

The main components of a radar are the transmitter, antenna, and receiver. The transmitter generates short pulses of energy in the radio-frequency portion of the electromagnetic spectrum. These are focused by the antenna into a narrow beam. They propagate outwards at essentially the speed of light. If the pulses intercept an object with different refractive characteristics from air, a current is induced in the object which perturbs the pulse and causes some of the energy to be scattered. Part of the scattered energy will generally be directed back toward the antenna, and if this back-scattered component is sufficiently large it will be detected by the receiver.

The primary function of radar is to measure the range and bearing of backscattering objects or "targets". Ranging is accomplished by a timing circuit that counts time between the transmission of a pulse and the reception of a signal. Direction is determined by noting the antenna azimuth and elevation at the instant the signal is received.

The fundamental radar display is the A-scope, an oscillograph trace of returned signal amplitude versus time after pulse transmission, as sketched in Fig. 10.1. Because the energy travels with velocity c, the time interval t between transmission and reception is related to target range by $r = ct/2$. The factor $1/2$ arises because the energy must make a round trip to range r in time t.

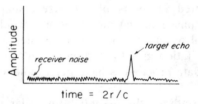

FIG. 10.1. Radar A-scan (schematic).

Timing begins from the initial transmission of each pulse, which under some conditions can lead to ambiguous range determinations. Suppose that a target is located so far from the transmitter that the return from a particular pulse is not received until after another pulse has been transmitted. In this case an erroneously close range is indicated. For a given radar pulse repetition frequency (PRF) there is a maximum range within which targets will be correctly indicated. Targets beyond r_{max} that return enough energy to be detected will be displayed ambiguously within this same range. This maximum unambiguous range is given by $r_{max} = c/2f_r$, where f_r denotes the PRF.

Some of the more important radar parameters, and their values for typical weather radars, are as follows:

1. Peak power (the instantaneous power in a pulse), P_t,

$$10 < P_t < 5 \times 10^3 \, \text{kw}.$$

2. Radio frequency, ν,

$$3 < \nu < 30 \, \text{GHz}$$

(corresponding to wavelengths between 1 and 10 cm).
3. PRF, f_r,

$$200 < f_r < 2{,}000 \, \text{sec}^{-1}.$$

4. Pulse duration, τ,

$$0.1 < \tau < 5 \, \mu\text{sec}.$$

An additional radar parameter of importance in meteorological work is the beamwidth, which is determined by the wavelength and the antenna size and shape. Beamwidth is defined with reference to the antenna pattern, which is a plot of the radiated intensity as a function of angular distance from beam axis. Such patterns will generally be different for planes through the axis having different orientations. Many antennas used with weather radars are paraboloidal, however, and the beam patterns are about the same for all planes through the axis. The beamwidth is usually defined as the angular separation between points where the transmitted intensity has fallen to half its maximum value, or 3 dB below the maximum.

Antenna patterns are characterized by sidelobes which are in general undesirable but unavoidable. Antenna design is concerned with achieving satisfactory compromises between characteristics of the mainlobe and the sidelobes.

The radar equation

By measuring the power returned by a target, one can often infer useful information about its nature. The basis of such inferences is the radar range equation, which relates the received power to the backscatter cross-section of the target. We consider first the case of a point target.

Suppose the radar transmits a peak power P_t. If this were radiated isotropically, a small area A_t at range r would intercept an amount of power given by

$$P_\sigma = \frac{P_t A_t}{4\pi r^2}.$$

The antenna is used to focus the energy in a narrow beam, increasing the power relative to the isotropic-radiated value. Allowing for the focusing effect, the small area A_t intercepts an amount of power given by

$$P_\sigma = G\frac{P_t A_t}{4\pi r^2},$$

where G is a dimensionless number called the antenna axial gain.

Now, if this area were to scatter the incident radiation isotropically, the power that would be returned to an antenna with aperture area A_e would be

$$P_r = \frac{P_\sigma A_e}{4\pi r^2} = \frac{GP_t A_t A_e}{(4\pi r^2)^2}.$$

The gain and the antenna aperture are approximately related by

$$G = \frac{4\pi A_e}{\lambda^2}.$$

It follows that

$$P_r = P_t \frac{G^2\lambda^2}{(4\pi)^3 r^4} A_t.$$

Most targets do not scatter isotropically, however, and as a convenient artifice the backscatter cross-section σ of the target is introduced, such that

$$P_r = P_t \frac{G^2\lambda^2}{(4\pi)^3 r^4} \sigma. \tag{10.1}$$

This is the form of the radar equation for a single target of backscatter cross-section σ. (Note that in general $\sigma \neq A_t$.)

The weather radar equation

Raindrops, snowflakes, and cloud droplets are examples of an important class of radar targets known as distributed targets. Such

targets are characterized by the presence of many effective scattering elements that are simultaneously illuminated. The volume containing those particles that are simultaneously illuminated is the
resolution volume of the radar, and is determined by beamwidth and
pulse length. For distributed targets whose scattering elements
move relative to each other, the power returned from a given range
is observed to fluctuate in time. Such fluctuations occur in weather
radar signals because the scatterers move relative to one another
owing to different fall speeds and wind variations across the
resolution volume. The instantaneous power of the fluctuating signal
depends upon the arrangement of the scatterers at that time and is
not simply related to their backscatter cross-sections. It turns out,
however, that a suitably long time average (in practice about
10 msec) of the received power from a given range is given by

$$\bar{P}_r = P_t \frac{G^2 \lambda^2}{(4\pi)^3 r^4} \sum \sigma, \tag{10.2}$$

where $\sum \sigma$ is the sum of the backscatter cross-sections of all the
particles within the resolution volume. This "contributing volume"
is given approximately by

$$V = \pi \left(\frac{r\theta}{2}\right)^2 \frac{h}{2}, \tag{10.3}$$

where $h = c\tau$ is the pulse length and θ is the beamwidth.
Sometimes (10.2) and (10.3) are combined to give

$$\bar{P}_r = P_t \frac{G^2 \lambda^2}{(4\pi)^3 r^4} \pi \left(\frac{r\theta}{2}\right)^2 \frac{h}{2} \eta, \tag{10.4}$$

where η denotes the radar reflectivity per unit volume.
Both (10.2) and (10.3) (and consequently 10.4) assume that the
antenna gain is uniform within its 3-dB limits, which is not true. In
(10.2) an average gain, somewhat less than the axial gain, should be
employed; also the effective volume could be defined as an integral
over the beam pattern instead of simply as the region within 3-dB
beam limits. On the assumption of a Gaussian beam pattern, (10.4) in
more accurate form becomes

$$\bar{P}_r = P_t \frac{G^2 \lambda^2 \theta^2 h}{1024 \pi^2 \ln 2} \frac{\eta}{r^2}. \tag{10.5}$$

This differs by only a factor $1/(2 \ln 2)$ from (10.4).

For a single spherical scatterer that is small compared to the radar wavelength (about 0.1λ is small enough), the backscatter cross-section is related to the sphere radius r_0 by

$$\sigma = 64 \frac{\pi^5}{\lambda^4} |K|^2 r_0^6, \tag{10.6}$$

where $K = (m^2 - 1)/(m^2 + 2)$ and $m = n - ik$ is the complex index of refraction of the sphere, with n = refractive index and k = absorption coefficient. This is called the Rayleigh scattering law; particles small enough for it to apply are called Rayleigh scatterers. The refraction term K depends on temperature and wavelength as well as the composition of the sphere. For the wavelengths employed in weather radars, and over the meteorological range of temperatures, $|K|^2 \approx 0.93$ for water and 0.21 for ice. Therefore an ice sphere has a radar cross-section only about 2/9, or 6.5 dB less, than that of a water sphere of the same size.

For a collection of spherical raindrops small compared to the wavelength, which are shuffling about, the average received power is

$$\bar{P}_r = P_t \frac{G^2 \lambda^2}{(4\pi)^3 r^4} 64 \frac{\pi^5}{\lambda^4} |K|^2 \sum r_0^6,$$

where $\sum$ again is a summation over the contributing volume. In terms of the raindrop diameters

$$\bar{P}_r = P_t \frac{G^2 \pi^5}{(4\pi)^3 r^4 \lambda^2} |K|^2 \sum D^6.$$

Thus for spherical scatterers small with respect to wavelength, the mean power received is determined by radar parameters, range, and by only two factors that depend upon the scatterers: the value of $|K|^2$ and the quantity $\sum D^6$. Because of the significance of the latter factor a new quantity Z is introduced, defined by

$$Z = \sum_v D^6 = \int N(D) D^6 dD, \tag{10.7}$$

where $\sum_v$ denotes a summation over *unit* volume and $N(D) dD$ is the number of scatterers per unit volume with diameters in dD. For raindrops $N(D)$ is the drop-size distribution. For snowflakes $N(D)$

is the distribution of melted diameters. (If this convention were not adopted for snow, the density of the snow would have to appear as a correction to $|K|^2$.)

In terms of Z, the radar equation, including the small correction for a Gaussian beam pattern, becomes

$$\bar{P}_r = \frac{\pi^3 c}{1024 \ln 2} \underbrace{\left[\frac{P_t \tau G^2 \theta^2}{\lambda^2}\right]}_{\text{RADAR}} \underbrace{\left[|K|^2 \frac{Z}{r^2}\right]}_{\text{TARGET}}. \qquad (10.8)$$

This is the most useful form of the radar equation, and the radar parameters are shown separate from the target parameters.*

Following (10.8), the received power may be related to the reflectivity factor Z by

$$10 \log \bar{P}_r = 10 \log Z - 20 \log r + C, \qquad (10.9)$$

where C is a constant—something like a sensitivity factor—determined by radar parameters and the dielectric character of the target. In this logarithmic form of the equation the power in decibels is related to the reflectivity factor as measured on a decibel scale. Usual conventions are that $\bar{P}_r$ is measured in milliwatts, with the quantity $10 \log \bar{P}_r$ called the power in dBm (decibels relative to a milliwatt), and Z is measured in mm^6/m^3 with the quantity $10 \log Z$ called the reflectivity factor in dBz. The logarithmic version of the equation is useful because of the wide ranges over which $\bar{P}_r$ and Z vary.

Relation of Z to precipitation rate

From its defining equation (10.7), Z depends on the drop-size distribution and is very sensitive to the large-drop component of the distribution. For a Marshall–Palmer distribution of raindrops extending from zero diameter to infinity, the reflectivity factor is given by

$$Z = N_0 \frac{6!}{\Lambda^7} = N_0 \frac{6!}{(41)^7} R^{1.47}.$$

---

* This is the form of the weather radar equation as given in unpublished notes on radar meteorology by Dr. P. L. Smith, South Dakota School of Mines and Technology.

This is in fair agreement with empirical data on Z and R for rain, which show that generally

$$Z = 200R^{1.6} \qquad (10.10)$$

to a reasonable approximation. Table 10.1 gives examples of Z values for several rainfall rates based on (10.10).

TABLE 10.1. *Reflectivity as a Function of Rain-fall Rate*

R (mm/hr)	0.1	1	10	100
Z (mm^6/m^3)	5	200	7950	316,000
dBz	7	23	39	55

A fundamental radar limitation is the noise level of the receiver. Without special provision, the received signal is not detectable unless it is stronger than the noise. Well-designed receivers have noise levels of about -105 to -110 dBm. For typical weather radars the values of the sensitivity factor C are such that the minimum detectable rainfall rate for a range of about 10 miles is in the order of 0.1 mm/hr, corresponding to drizzle. Consequently, weather radars usually detect rain but not cloud. For cloud studies, special radars with wavelengths of about a centimeter are employed.

There is more variability among the Z–R relations for snow than for rain, but an approximate relation that is generally accepted is

$$Z = 2000R^2, \qquad (10.11)$$

where, as in (10.10), R denotes precipitation rate in mm of water per hour.

Radar displays and special techniques

The most common display is the PPI (Plan Position Indicator) which maps the received signals or "echoes" on polar coordinates in plan view. With elevation angle fixed, the antenna scans 360° in azimuth with the beam sweeping across a conical surface in space. At every azimuth the voltage output of the receiver as a function of range is used to intensity-modulate a tube with polar coordinates. The distribution of precipitation in plan view is thereby produced, and a time sequence of PPI's indicates the development and motion of precipitation areas. One full azimuth scan requires in the order of 10 sec and photographic records are usually kept at the rate of one frame per revolution.

Without careful calibration and maintenance procedures, PPI records do little more than show where and when precipitation is occurring and indicate roughly where the rain is relatively intense (bright echoes). This information in itself is useful in synoptic meteorology and in cloud physics investigations, but is of limited value in quantitative precipitation studies. For this work, it is desirable to know the actual distribution of Z within the echoes. In principle this information can be obtained from PPI film records by densitometry or some equivalent form of exposure analysis. In practice it is extremely difficult to maintain the required overall system calibration—from radar receiver to film processing—for this approach to be accurate.

Instead of this approach, a thresholding amplifier is often employed in the receiver which converts the continuous range of received signal power to a stepped scale of up to 6 or 7 levels. These then appear on the PPI as different shades of gray, and about 5 can be readily resolved if only ordinary darkroom procedures are followed. Generally the steps are arranged such that they correspond to 10-dB intervals in reflectivity. A gray-shade PPI thus will typically allow the estimation of target reflectivity factor to within ±5 dBz.

Similar to the PPI, the RHI (Range Height Indicator) is a display which is generated when the antenna scans in elevation with azimuth fixed. While the PPI emphasizes horizontal echo structure, the RHI shows vertical structure in detail.

Ideally, one would like three-dimensional echo coverage, which neither the PPI nor RHI provides. This can be achieved by a programmed antenna scan, in which azimuth and elevation are systematically varied to survey all or most of space around the radar site. Analog techniques have been used to combine the data from a spiral scan automatically, to produce constant-altitude PPI (CAPPI) maps at several levels above the ground. Digital automatic procedures have also been used to provide vertical or horizontal echo cross-sections over limited regions of the total area surveyed. All such techniques employing a three-dimensional scan require more time than the simple PPI and RHI—usually about 5 min. The more modern digital techniques have the advantage of working directly with the received signals, bypassing the photographic step altogether.

In modern radar sets, it is not uncommon for the transmitter to be coherent. By this it is meant that the frequency of the transmitted signal is constant and that each pulse bears the same phase relation to its predecessor. This is the kind of signal produced by pulse-modulating a free-running stable oscillator; it is a characteristic of radars having klystron transmitters. Using such equipment, and making special provisions in the receiver, it is possible to make Doppler velocity measurements. In effect, the frequency content of the returned signal is compared with that of the transmitted signal, and frequency shifts are interpreted as arising from the Doppler effect. Thus a frequency shift $\Delta \nu$ corresponds to a velocity $\vec{V}$ according to

$$\Delta \nu = \frac{2}{\lambda} \vec{V} \cdot \hat{r},$$

where $\hat{r}$ denotes a unit vector in the radar-pointing direction.

Meteorological targets induce a spectrum of Doppler shifts because their scattering elements generally do not all move with the same velocity. Most of the work in Doppler radar studies is concerned with meteorological interpretations of the Doppler spectrum. When the beam is pointed vertically, the Doppler spectrum contains information about vertical air motions and precipitation fall speeds. For horizontal viewing, Doppler velocities are interpreted as arising from horizontal air motions. The scattering particles move with the wind to a close approximation, though the echo systems frequently do not move exactly with the wind because of precipitation development or dissipation in preferred regions of the echo.

To measure target reflectivity, the amplitude of the returned signal is compared to that of the transmitted signal. In Doppler velocity measurements, the frequencies of returned and transmitted signals are compared. It is also possible to derive information about the target by comparing the polarization of the received and transmitted waves. Non-symmetrical scattering objects induce an amount of cross-polarization depending in a complex way on their shapes, sizes (compared to a wavelength), and dielectric properties. For precipitation, a theory has been developed which relates the cross polarization to the axial ratio of the particles, which are approximated as ellipsoids small compared to the wavelength. There is some indication that polarization techniques provide a method of distinguishing between rain and other precipitation forms.

Problems

1. Consider the following model raindrop population:

(i) Drop-size distribution given by

$$N(D) = N_0 \exp(-bD), \quad 0 \le D \le \infty,$$

where $N_0 = 0.08 \text{ cm}^{-4}$, a constant, and b is a parameter that depends on rainfall rate.

(ii) Fall speed law approximated as

$$u(D) = kD,$$

where $k = 4 \times 10^3 \text{ sec}^{-1}$.

For this model calculate the relation between radar reflectivity factor Z (mm^6/m^3) and rainfall rate R (mm hr^{-1}).

2. A particular radar observation in steady-state, stratiform rain shows a strong decrease of reflectivity with height, with a 20 dB change in the lower 1 km of cloud. Estimate the value of the product EM in this cloud layer, making the following assumptions:

(1) All raindrops at a given level are of the same size.
(2) Raindrop growth is described by the elementary form of the continuous-growth equation for accretion, with E the effective collection efficiency and M the cloud liquid water content, assumed constant.
(3) The raindrops at cloud base have a diameter of 1 mm.
(4) The same linear fall speed law applies as given in problem 1.
(5) Vertical air motion is negligible.

3. At a time early in the development of a cumulus congestus cloud the radar reflectivity factor equals -30 dBz and the droplet spectrum has a Gaussian shape, centered at radius 8μm and with a dispersion $\sigma/\bar{r}$ of 0.15. Assume that these droplets grow only by condensation, with the supersaturation constant at 0.5%. Solve for the reflectivity factor (in dBz) at 5, 10, and 15 min later.

Use the classical form of the condensation-growth equation, neglecting a/r and b/r^3. As a further approximation, assume the process takes place at 700 mb and 0°C.

4. The following model is found to be satisfactory for explaining certain radar observations in Hawaiian orographic rain:

(i) Liquid water content L, in the form of raindrops, is constant throughout the raincloud.

(ii) At a given altitude all raindrops are of the same size.

(iii) The size of raindrops increases with distance fallen due to coalescence. This causes a reduction in number density given by:

$$\frac{dN}{dz} = KN,$$

where N is the number density of raindrops at height z and K is a constant.

Derive an expression for the height-dependence of the radar reflectivity factor Z in this model, assuming a steady-state condition, that is, no time variations of cloud properties at a given altitude. Show that the decrease of reflectivity with altitude over the vertical distance Δh is given by $4.343 \, K\Delta h$ decibels.

CHAPTER 11

PRECIPITATION PROCESSES

THE areal extent, intensity, and lifetime of a precipitation system are largely controlled by vertical air motions. Accordingly, it is customary to classify precipitation as one of two types, depending on the dominant mechanism responsible for the vertical motion:

1. Widespread, stratiform, continuous precipitation associated with large scale ascent produced by frontal or topographic lifting or large scale horizontal convergence.
2. Localized, convective, showery precipitation associated with cumulus-scale convection in unstable air.

This is a useful classification, although the distinction between stratiform and convective precipitation is not always sharp. Widespread precipitation, when observed either by radar or raingauge, invariably shows fine-scale structure with the most intense precipitation confined to elements with a size of only several kilometers. Precipitation of convective origin can extend over a large area and produce a pattern similar to that of continuous precipitation. Nevertheless, it is usually possible to describe a pattern as either markedly nonuniform (hence convective), with locally intense regions ranging in size from 1 to 10 km and separated from one another by areas free of precipitation, or rather uniform (hence stratiform) with less pronounced small scale structure and a wider overall extent. Moreover, the pattern of stratiform precipitation evolves relatively slowly in time, and that of convective precipitation changes rapidly.

Stratiform rain is produced in nimbostratus clouds, although dissipating cumulus clouds and orographic clouds may contain rain with stratiform structure. Most snow originates in nimbostratus clouds, but snow flurries and graupel showers can be produced in convective clouds.

Widespread precipitation

Figure 11.1 is an example of the radar pattern of stratiform rain. Although reflectivity variations are present, including several small regions of heavy rain and a band oriented approximately NE–SW, the pattern is relatively uniform over large distances. In this kind of rain vertical air motions are weak and gravitational settling determines much of the pattern. Consequently the vertical structure of the pattern is closely related to the precipitation growth process.

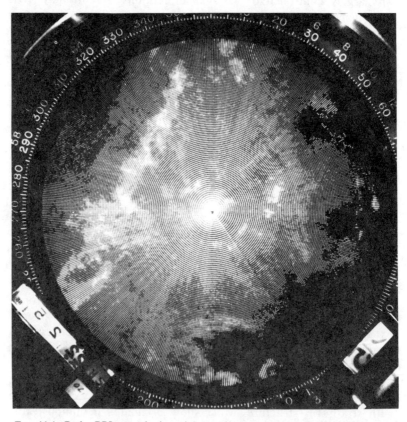

FIG. 11.1. Radar PPI map of rain mainly stratiform in structure. Maximum range 40 miles. Display consists of an array of discrete dots whose brightness and size are proportional to the reflectivity factor, in steps of 10 dB. (From McGill Radar Weather Observatory.)

Figure 11.2 is an example of the vertical structure of a pattern of light rain dominated by gravitational settling. The observations were obtained by a Doppler radar with a vertically pointing beam. As the rain pattern moved through the fixed beam, the signal intensity and Doppler velocity were measured continuously as functions of altitude. In coordinates of height versus time, the data portray the vertical rain structure in detail. The observed time variations arise from the translation across the beam of spatial variations in the rain and from actual variations of the rain structure in time. In this example the pattern had a velocity of about 3 m/sec. Therefore the record of 24 min duration corresponds to a distance through the rain pattern of about 4 km. Since the reflectivity contours are approximately horizontal, especially during the latter half of the record, it is reasonable to suppose that the rain pattern is slowly changing and that the figure approximately represents the rain pattern in space, along a line in the direction of its motion.

The horizontally stratified pattern, with reflectivity and downward

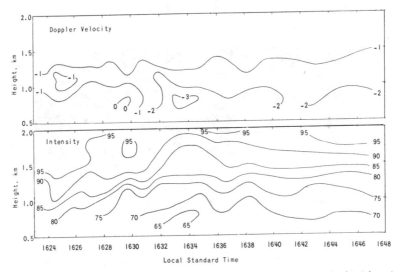

FIG. 11.2. Time-height patterns of reflectivity (below) and Doppler velocity (above) in Hawaiian orographic rain. Intensity values I are related to reflectivity factor in dBz by the formula $dBz = 95 - I$. The contour $I = 65$ thus corresponds to 30 dBz, and $I = 95$ corresponds to 0 dBz. Doppler velocites in m/sec, with negative signs indicating downward motion (toward the radar). (From Rogers, 1967.)

Doppler velocity increasing progressively with distance downward from the echo top, indicates a steady precipitation process in which small raindrops near the echo top slowly descend through cloud and small raindrops, growing by coalescence. Such a process results in an increase in average drop size with distance fallen, with corresponding increases in the reflectivity.

The best examples of relatively uniform, widespread precipitation are found in snow. Figure 11.3 shows vertical cross-sections through snow at four different times on the same day. The pictures at 1800 and 2012 are examples of the smooth, stratiform structure often observed in snow.

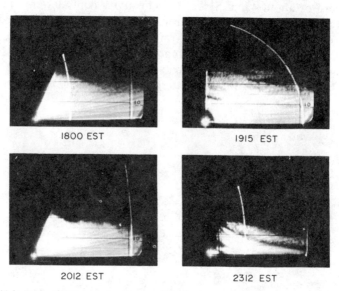

1800 EST   1915 EST

2012 EST   2312 EST

FIG. 11.3. Radar RHI records of snow. Maximum range 10 miles for the picture at 1915, 25 miles for the rest. Altitude marks at 5000 ft intervals. (From Wexler and Austin, 1954.)

When it is warm enough for snow to melt before reaching the ground, there is often observed a thin layer of relatively high reflectivity just below the level of 0°C. This radar "bright band" is the melting layer, the region of transition from snow to rain (Fig. 11.4). As snowflakes descend into the melting layer their radar

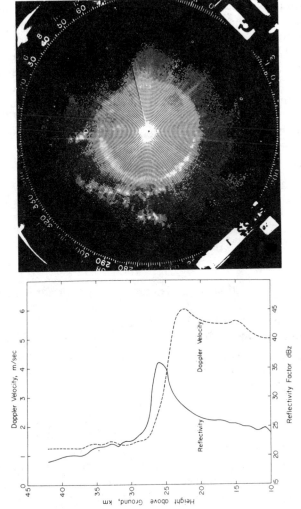

FIG. 11.4. Two views of the radar bright band: at the left a vertical profile of reflectivity and Doppler velocity as measured with a vertically pointing Doppler radar; at the right a PPI map at 8° elevation on which the melting layer appears as a bright ring at a range of about 12 miles. (Doppler data from Cornell Aeronautical Laboratory; photograph from McGill Radar Weather Observatory.)

reflectivity increases for various reasons, the most important of which is melting because the dielectric constant of water exceeds that of ice by a factor of four (6 dB). Also, in its initial stages, melting produces distorted wet snowflakes with somewhat higher reflectivities than those of spherical drops of the same mass. Continuing to melt while descending, the snowflakes become more compact and finally collapse into raindrops. Since the raindrops fall faster than the snowflakes their concentration in space is reduced. This dilution of the numbers accounts in part for the decrease of reflectivity in the lower part of the melting layer. If the melting flakes break apart further reduction in reflectivity occurs.

Wexler (1955) analyzed the processes in the melting layer and gave the following estimates of the changes in reflectivity arising from the different effects:

	Melting	Fall velocity	Shape	Condensation	Total
Snow to bright band	$+6$	-1	$+1\frac{1}{2}$	0	$+6\frac{1}{2}$ dB
Bright band to rain	$+1$	-6	$-1\frac{1}{2}$	$+\frac{1}{2}$	-6 dB

The fact that observations often reveal a stronger bright band than predicted suggests that aggregation in the upper part of the melting layer and disintegration below are occurring.

There are occasions in widespread precipitation when the bright band is weak, diffuse or entirely absent because of convective overturning: mixing disrupts the stratification necessary for the melting layer to be well defined.

Convection in widespread precipitation also manifests itself in so-called snow "generating cells". Figure 11.5 shows four examples of the structure of snow in the vertical, obtained with a vertically pointing radar. The snow is widespread in every case, but cases (b) and (c) are more stratiform in appearance than the others. Patterns of type (a) and (d) are observed in advance of snow reaching the ground; most snowfall is associated with a trail pattern as in (b). These trails originate in compact cells called generating cells which are smaller than the trails and not often actually observed. Pendulous extensions of the lower edge of the echo in (d) are referred to as

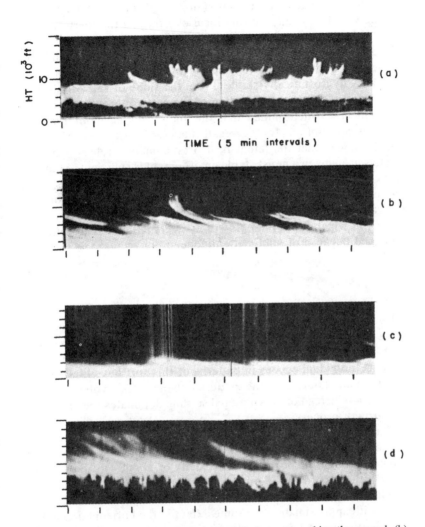

FIG. 11.5. Time-height records of (a) snow aloft, but not reaching the ground, (b) well-defined trail pattern, (c) relatively homogeneous echo, (d) stalactites (downward protrusions) at leading edge of storm. Total length of records is 50 min; vertical extent is 20,000 ft. (From Douglas *et al.*, 1957.)

stalactites, and occur when snow falls into dry air. Sublimation of the snow chills the air, causing local overturning that perturbs the lower echo boundary.

Snow generating cells are found to have no preferred altitude or temperature, but to be located usually just above frontal surfaces in air that is hydrostatically stable. From the slope of snow trails and knowledge of the speed of motion of the pattern through the beam, Marshall (1953) determined that the snow falls at about 1 m/sec, which is appropriate for aggregate flakes. Douglas and Marshall (1954) showed that the latent heat of sublimation released by ice crystals growing in a moist stable environment is sufficient to initiate convective overturning, and that the vertical development of these convective elements is comparable to the observed dimensions of generating cells. Consequently the generating cells are probably regions in which ice crystal growth by aggregation and accretion is enhanced because of sublimation-induced convection.

Showers

Two examples of PPI records of showers are shown in Fig. 11.6. Some of the showers in the first example are arranged in a line to the west of the radar, but in the second no particular organization is evident. Individual echoes in patterns of this sort have lifetimes of less than an hour, and the patterns themselves evolve rapidly. Convection instead of gravitational settling dominates the precipitation growth process.

Figure 11.7 shows the vertical structure of a rainshower as measured by a vertically pointing Doppler radar. The shower was moving through the beam with a velocity of about 4 m/sec. Consequently the record, of 18-min duration, corresponds to a distance through the shower of about 4 km. The shower is thus of compact form, with approximately the same extent in the horizontal as in the vertical. Figure 11.7 does not depict the exact form of the shower in space, because development and internal changes were undoubtedly occurring during the observation time.

The Doppler velocity pattern of this shower (not shown) was used in connection with the reflectivity data to deduce the pattern of updraft velocity in Fig. 11.7b. At 1414 there was upward air motion

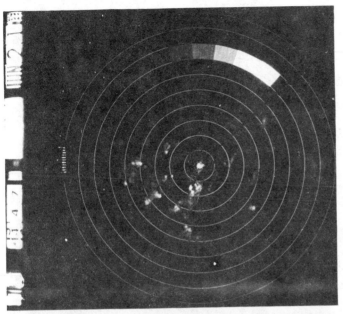

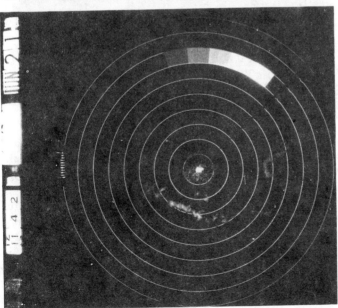

Fig. 11.6. Two examples of radar records of showers. In the picture on the left some echoes are organized in a line; the echoes in the right picture are located randomly. Range rings at 10-mi intervals. Grey scale thresholds in 10-dB steps, with calibration pattern at 70 mi to the east. (From Alberta Hail Studies Laboratory.)

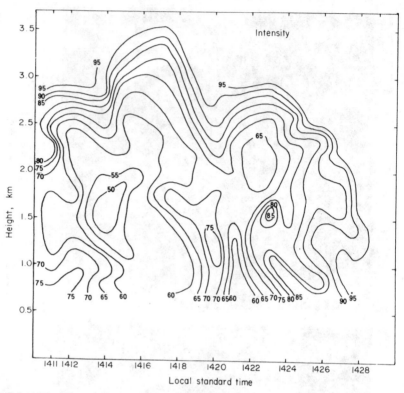

FIG. 11.7a. Time-height pattern of reflectivity in a warm-rain shower observed in Hawaii. Intensity values I are related to reflectivity factor in dBz by the formula $dBz = 95 - I$.

through the entire vertical extent of the pattern, with maximum velocities exceeding 5 m/sec. These upward motions include the 45 dBz maximum in the reflectivity pattern. At 1421 the signal again intensified overhead in connection with a region of vertical air motion, this time somewhat weaker than earlier. Heavy rain fell while the shower passed; total accumulation during the 18 min period was 3.8 mm.

The broad features of this example of a Hawaiian shower are relatively simple. It appears to consist of two active convective elements, each having a fairly continuous updraft and an associated region of high reflectivity. Most showers that have been observed in

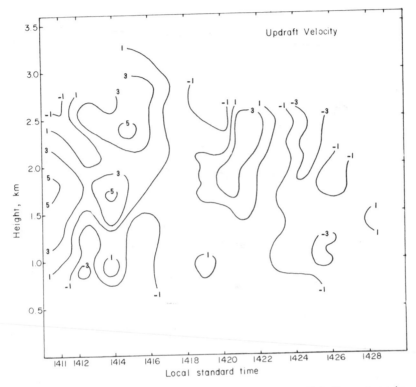

FIG. 11.7b. Pattern of updraft velocity in the same shower. Velocity contours in m/sec, positive upwards. (From Rogers, 1967.)

this manner appear to be multicellular. There are uncertainties in interpreting time-height records, however, for it is often not possible to determine whether the core of the shower or a fringe area is passing overhead.

Browning *et al.* (1968) reported on a shower which was observed simultaneously by two Doppler radars, one pointing vertically and the other at a low elevation angle. The center of the shower was known to pass directly over one of the radars during the observing period, and from the data it was possible to infer the pattern of air motion in the shower and the region of precipitation growth (Fig. 11.8). The updrafts are confined to the upper part of the shower

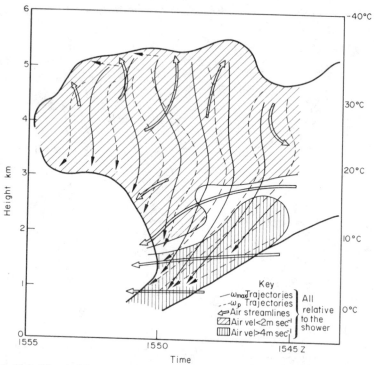

FIG. 11.8. Time-height pattern of a shower showing streamlines of air motion, trajectories of the largest particles (ω_{max}), and trajectories of the particles contributing most to the Doppler velocity (ω_p). The streamlines and trajectories are drawn relative to the shower. Temperature profile indicated on the right. (From Browning et al., 1968.)

(an observation in common with many other Doppler radar studies of showers) but are very weak, amounting to only about 1 m/sec. Precipitation in the form of graupel is thought to originate and begin growing in this area. The graupel continues to grow as it descends through the weak updrafts and forms a precipitation streamer at lower levels. The solid trajectories in the figure are estimates of the paths of the largest graupel particles. Downdrafts predominate at low levels, again in common with other radar-measured airflow patterns. This rather characteristic pattern—updrafts at high levels and downdrafts below—suggests that the initial ascent of the air from low levels takes place outside any existing precipitation and is completed by the time the cloud particles in the ascending air have grown to radar detectable size. The lifetime of the convective element is thus about the same as the time required for the precipitation to develop. This fundamental characteristic of showers was pointed out by Houghton (1968), and will be mentioned later in the context of cloud modification.

Precipitation theories

Precipitation from stratiform clouds is thought to develop principally by the ice-crystal process. These clouds have relatively low liquid water contents, so coalescence is likely to be ineffective. The clouds last a long time, however, and if cloud persists at altitudes where the temperature is about $-15°C$ the ice-crystal process can lead to precipitation. As explained by Braham (1968) in a survey of precipitation development, each level in stratiform clouds has a special role to play in the precipitation process. The cold upper levels ($T \approx -20°C$) supply ice crystals that serve as embryos for precipitation development at lower levels. The cloud at midlevels ($T \approx -15°C$) provides the right environment for rapid diffusional growth. Aggregation and accretion proceed most rapidly still lower in the cloud, at temperatures between $-10°C$ and $0°C$. Most of the precipitation growth occurs in these lowest levels.

In convective clouds less time is available for precipitation growth, but since liquid water contents are typically higher than in stratiform clouds coalescence stands a better chance of producing rain. From the observation that the lifetime of a convective element (about 20 min) is also the time needed for precipitation to grow, Houghton (1968) concluded that the precipitation forming process must begin early in the developing cloud and therefore at a low level. While the precipitation may be initiated by coalescence or the ice-crystal process, depending primarily on the temperature and cloud water content, most precipitation growth is by accretion.

Thus the mechanisms of precipitation formation are quite different in stratiform and convective clouds. As a useful approximation continuous rain can often be viewed as a steady-state process, in which cloud quantities may vary with height but are constant with time at any given height. Conversely, showers may be approximated as systems in which the cloud properties vary with time but are

constant with height at any given time. These limiting approximations were first suggested by Rigby, Marshall, and Hitschfeld (1954).

As an example of the use of the approximation for showers we
now solve for the evolution of a raindrop-size distribution with time,
assuming growth by accretion of cloud droplets. In this case, the
elementary form of the continuous growth equation is given by
(7.15),

$$\frac{dR}{dt} = \frac{\bar{E}M}{4\rho_L} u(R).$$

For drops in the intermediate size range the linear fall speed law
(7.8) applies, and if $\bar{E}M$ may be regarded as constant the solution of
the growth equation is

$$R(t) = R(0)e^{at}, \tag{11.1}$$

where $a = k_3\bar{E}M/4\rho_L$.

Now let $n(R,t)$ and $n(R,0) = n_0(R)$ denote, respectively, the raindrop-size spectrum at time t and at the initial time. Since accretion is
the only growth mechanism considered the number of raindrops in
the interval dR_0 in the initial distribution is the same as the number
of drops in the interval dR of the distribution at time t. That is,

$$n(R,t)dR = n_0(R_0)dR_0, \tag{11.2}$$

which is analogous to (6.29) for growth by condensation. From
(11.1),

$$n(R,t) = e^{-at}n_0(Re^{-at}), \tag{11.3}$$

which is the solution we sought, expressing the distribution at any
time t in terms of the initial distribution. This approximation neglects
coalescence among raindrops, and is probably most appropriate for
the early stages of raindrop growth in convective clouds, when light
rain is falling through relatively dense cloud.

In the continuous-rain approximation the number flux of raindrops is constant with height; otherwise the drop-size distribution
would vary with time. Therefore the initial distribution and the
distribution after distance of fall h are related by

$$n(R,h)u(R)dR = n_0(R_0)u(R_0)dR_0. \tag{11.4}$$

Assuming negligible updraft speed and again employing (7.8) for the
fall speed, we find

$$n(R,h) = \left(1 - \frac{bh}{R}\right)n_0(R - bh),\qquad(11.5)$$

where $b = \bar{E}M/4\rho_L$.

Though only coarse approximations, (11.3) and (11.5) give an indication of the difference between the two idealized precipitation processes. In effect, these results represent extensions of the continuous-growth accretion equations to drop populations.

Mesoscale structure of rain

Details of precipitation patterns and their causes have come under study only fairly recently. Browning and Harrold (1969) analyzed the air motion and precipitation patterns associated with an occluding cyclone that passed over England and Wales, and presented the schematic distribution of rainfall types shown in Fig. 11.9. Rain relatively uniform in pattern was located well ahead of the surface warm front, and there was a fairly distinct transition to showery rain about 150 km from the front. Bands of showers in the warm sector

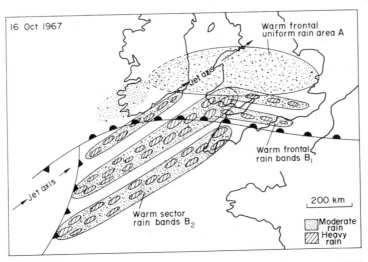

FIG. 11.9. Schematic diagram showing the distribution and structure of rainfall in a cyclone wave. The axis of the high level jet stream is shown passing directly over the Irish Sea. (From Browning and Harrold, 1969.)

were orographically influenced and aligned parallel to the winds at about 700 mb. These bands extended ahead of the warm front, merging with smaller bands aligned with the front. The low-level air ahead of the warm front was convectively unstable. The changing character of the rain was a result of ascent at the surface warm front, which eventually triggered the instability.

Using radar observations and raingauge records, Austin and Houze (1972) studied the precipitation patterns of nine New England storms covering a wide range of seasonal and synoptic situations. Although at first glance quite dissimilar, all patterns were found to be composed of subsynoptic-scale precipitation areas with rather clearly definable characteristics. These areas could be grouped into four categories: synoptic areas which are larger than 10^4 km^2 and have a lifetime of one day or longer; large mesoscale areas which range from 10^3 to 10^4 km^2 and last several hours; small mesoscale areas which cover 100–400 km^2 with a lifetime of about an hour; and smaller elements which are about 10 km^2 in size and last usually no longer than half an hour. In all cases studied it was found that every precipitation area of any of these scales contained one or several of each of the smaller sized areas. An investigation of the precipitation intensities within the various areas showed that the rainfall rates in large mesoscale areas were 2–4 times greater than those on the synoptic scale; that the rain rates in the small mesoscale areas were about double those in large mesoscale areas; and that the rain rates in the smallest elements were 2–10 times those in the small mesoscale areas. Although the smallest elements have the highest rain rates, the main contribution to the total rainfall on the synoptic scale comes from the small and large mesoscale areas.

Summarizing studies of this kind, Harrold and Austin (1974) have noted that regions of heavy rain can be found in widespread rain as well as showery situations and tend to occur in compact groups rather than to be randomly scattered. The groups are often in the form of bands, typically about 20 km wide and in well organized cases several hundred kilometers long. The bands are usually related to frontal surfaces or squall lines, but need not be parallel to them. Topographic effects, as well as fronts, can affect the structure and development of rain areas. Although the effects have been documented and to some extent categorized, it is not yet understood why precipitation has the strong tendency to become organized into the characteristic scales and patterns that are observed.

A different approach to describing the fine-scale structure of rain has been taken by Zawadzki (1973) and others. The rate of rainfall R is a function of position on the surface (x,y) and time t. Following the practice in the theory of random processes, it is possible to define the autocorrelation function of rainfall rate in time or space. The space autocorrelation function, for example, indicates the extent to which the rainfall pattern is statistically uniform in space.

Zawadzki determined $R(x,y,t)$ from radar data and used an optical technique to form the autocorrelation functions. From measurements in widespread rain, he found that in any direction the space autocorrelation decays with distance approximately exponentially, with the correlation dropping to the value $1/e$ at distances between 20 and 35 km. The rate of decay of the autocorrelation with distance was about the same in all directions for distances less than 10 km, implying a statistically isotropic structure of precipitation in the small scales. The time autocorrelation function for the pattern studied was found to decay to the value $1/e$ in about a half hour, implying that this is approximately the time required for the fine-scale rain structure to evolve or reorganize into a different configuration.

Precipitation efficiency

Clouds provide the intermediate step in converting atmospheric water vapor to precipitation. Not all rainclouds are equally effective in accomplishing this conversion. Small cumulus clouds, for example, often grow rapidly but begin to dissipate just as precipitation develops. Consequently much of the cloud water is not converted to precipitation but remains aloft, eventually to evaporate. For a different reason many stratiform clouds are also ineffective in producing precipitation. Although they may last for hours, they have neither the high liquid water contents which favor coalescence nor the cold temperatures needed to initiate the ice crystal process. Therefore little precipitation occurs even though the cloud may be supercooled aloft, and hence microphysically unstable with respect to the ice crystal process. The concept of precipitation efficiency has been used to describe, from several points of view, how effectively a cloud converts either vapor or condensed material to precipitation.

Braham (1952) determined the water budget of small thunderstorms by analyzing extensive data on clouds in Florida and Ohio. For the average inflow of water vapor into these storms he found 8.9×10^8 kg. Of this amount, 5.3×10^8 kg is condensed; the remainder leaves the storm without condensing. Of the water that condenses, only about 10^8 kg reaches the ground as rain. The rest evaporates in the downdraft or at the sides of the cloud. The precipitation efficiency may be defined as the ratio of the mass of rain reaching the ground to the mass of vapor entering the cloud. With this definition the efficiency is only 11%. If the efficiency is defined as the fraction of condensed water that ultimately reaches the ground, it becomes 19%, a figure often quoted for thunderstorms.

Wexler (1960) defined the precipitation efficiency somewhat differently, as the ratio of the amount of precipitation that falls to the water made available by condensation in pseudoadiabatic ascent. He found that the most efficient clouds in this sense are cumulus embedded in widespread stratus.

Looking at precipitation efficiency from the microphysical point of view, Hardy (1963) made calculations with different forms of the raindrop size distribution $N(D)$ to determine which were the more effective in depleting (sweeping out) the cloud droplets. He concluded that steep distributions (large values of Λ in the exponential formula 9.1) were the most effective in this sense. On the basis of radar observations in Hawaiian orographic rain, which is characterized by numerous small drops, Rogers (1967) found support for Hardy's idea, noting that the exceptionally rapid decrease of reflectivity with height implies efficient sweepout.

The sweepout efficiency was examined by Houghton (1968) in an appraisal of schemes for artificial precipitation modification. He defined the efficiency S by

$$S = \frac{\pi}{4} \int N(D)D^2 V(D)dD, \tag{11.6}$$

where the integration extends over all drops, assumed spherical. Physically, S is the fraction of a unit horizontal area in the cloud that is geometrically swept out by precipitation particles in unit time.

From data on drop-size distributions, Houghton determined that to a good approximation $S \propto R$. He found that the "scavenging" is incomplete for rain-showers, which is one cause of their low precipitation efficiency (Braham's 19%). With regard to the feasibility of modifying precipitation, Houghton concluded that owing to the inefficiency of showers opportunities exist for enhancing rain by cloud seeding, but these opportunities occur only under certain specific conditions and at particular times.

Problems

1. The sweepout efficiency of a single precipitation particle may be defined as the volume of space that it geometrically sweeps out per unit time. Compare the sweepout efficiency of a graupel particle with that of a small raindrop having the same mass. For graupel, assume that mass and radius are related by

$$m = 0.52r^3 \quad \text{(c.g.s.)}$$

and that the terminal fall speed is given by

$$u(r) = 520r^{0.6} \quad \text{(c.g.s.).}$$

For the raindrop, assume the linear fall speed law

$$u(r) = kr,$$

with $k = 8 \times 10^3 \text{ sec}^{-1}$.

2.* The vertical gradient of reflectivity in stratiform rain gives an indication of the extent to which raindrops are growing by sweeping out cloud droplets. A strong decrease of Z with height (or increase with distance fallen) indicates rapid accretional growth.

   In the lowest one kilometer of a particular nimbostratus cloud the raindrops are growing by accreting cloud droplets. The effective cloud liquid water content (the product of E times M) equals 2 g/m³. Using the elementary form of the continuous growth equation, show that the reflectivity increases by approximately 19 dB in the lowest kilometer of the cloud. Solve also for the change in (rain water) liquid water content. Make the following assumptions:

   (1) steady-state process
   (2) zero updraft velocity
   (3) elementary form of continuous growth equation
   (4) drop growth by accretion only
   (5) raindrop size distribution at 1 km above cloud base of Marshall–Palmer form corresponding to $R = 0.1$ mm/hr.

3. In a rapidly developing cumulonimbus cloud, the radar echo is observed to appear simultaneously over a deep interval of height. For the particular radar used, this

---

*See footnote on p. 86.

initial echo corresponds to drizzle-sized drops. As the cloud continues to develop, these drops grow rapidly by sweeping out cloud droplets, and the radar echo strengthens at all levels.

Calculate the time required for the signal to increase by 10 dB. Make the following assumptions:

(1) Cloud liquid water content is constant at 5 g/m³.
(2) All the precipitation growth is due to sweepout of cloud droplets; there is no growth by diffusion or by coalescence among the raindrops.
(3) The effective average collection efficiency is 0.5.
(4) The raindrops are of relatively small size, so that their terminal fall speed is accurately approximated by $u(r) = kr$, where $k = 8 \times 10^3$ sec^{-1}.
(5) The raindrops account for all the reflectivity; the contribution of cloud to the signal is negligible.
(6) Breakup effects are negligible.

4. In the theory of the radar "bright band" one of the effects considered is the increase in reflectivity within the melting layer arising from condensation on the surface of the melting snowflakes. When snow falls into this layer it melts, chilling the air. This chilling will result in some condensation onto the surface of the melting particles. Show that approximately 60 g of water is condensed for each kilogram of snow that melts.

5. Develop an expression for the rate of depletion of cloud water content M as a result of collection by raindrops. As in Problem 2, p.146, assume a general negative-exponential form for the raindrop size distribution and a linear dependence of fall speed on drop diameter. Show that in a stagnant cloud, that is one in which there is no vertical air motion and no change in cloud water content except by rain-collection, rain falling at a steady rate of 10 mm/hr for 5 min will reduce M to approximately 45% of its initial value.

CHAPTER 12

SEVERE STORMS AND HAIL

IN air that is sufficiently moist and unstable, convective clouds can grow to great heights, develop vigorous updrafts, and produce heavy rain, lightning, and hail. These large severe storms may occur individually or, more typically, in groups associated with synoptic-scale fronts or mesoscale convergence areas. In many parts of the world they are the cause of serious flooding, wind and hail damage, and loss of life.

Although our understanding of such storms is incomplete, extensive studies over the past two decades by means of radar, radiosonde networks, and instrumented airplanes have made it possible to describe their structure and development and to recognize the meteorological conditions under which they are likely to occur. The main processes relevant to the development of severe storms were reviewed by Newton (1967). This chapter describes the structure of thunderstorms and outlines the theory of hail growth.

Life cycle of the thunderstorm cell

From extensive observations of thunderstorms in Florida and Ohio, Byers and Braham (1949) found that the storms are made up of one or more units of convective circulation, consisting of an updraft area and a region of compensating downward motion. These convective cells are much the same in structure and behavior in most storms and may therefore be considered as a class of convective phenomena unique to thunderstorms. Often a cloud is made up of a number of cells in various stages of development, and it is difficult to identify any individual cell. However, it is convenient to consider the thunderstorm cell as the elementary unit of storm structure.

The life cycle of a cell is divided into three stages depending on the predominant direction and magnitude of the vertical air motion:

(1) Cumulus stage—characterized by an updraft throughout most of the cell.

(2) Mature stage—characterized by the presence of downdrafts and updrafts.

(3) Dissipating stage—characterized by weak downdrafts throughout most of the cell.

These stages and the accompanying precipitation forms are indicated in Fig. 12.1.

As the updraft causes the cloud to grow in the cumulus stage, air flows in through the sides ("entraining") and mixes with the updraft. With continued upward motion a large amount of water condenses and eventually falls as precipitation. This falling water initiates the downdraft because of viscous drag of the water on the air and evaporative cooling of the air. This is the start of the mature stage of development. The air of the downdraft reaches the ground as a cold core in the rain area and spreads over the surface, changing the surface wind pattern.

The downdraft interferes with the updraft at low levels in the cloud, and eventually cuts off the updraft from its source region. The cell then enters its dissipating stage. With the decay of the updraft and consequent elimination of the source of rainfall the downdraft weakens and finally dies out completely, leaving a residue of cloudy air.

The cumulus stage typically has a duration of 10–15 min. The mature stage lasts 15–30 min; though difficult to specify definitely, the dissipating stage lasts about 30 min.

Since the time of Byers and Braham's study, long-term field programs to investigate severe storms have been mounted and more sophisticated techniques have been developed for the observations. It has been found nevertheless that the relatively simple model of Fig. 12.1 adequately describes the general features of small, "single-cell" cumulonimbus clouds and of the convective elements in larger "multicell" clouds. Figure 12.2, for example, shows the vertical air velocity measured by a Doppler radar during the early mature stage of a growing cumulonimbus. Upward velocities are consistent with

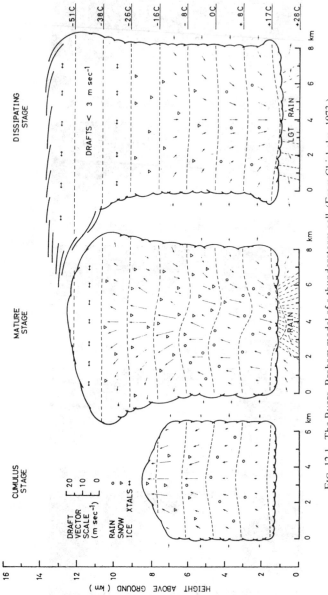

FIG. 12.1. The Byers–Braham model of a thunderstorm cell. (From Chisholm, 1973.)

1356 MST ↑

1401 MST ↑

1406 MST ↑

1411 MST ↑

1416 MST ↑

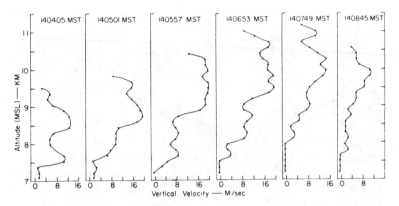

Fig. 12.2. A developing cumulonimbus cloud and radar-measured updraft velocities. The pictures indicate two cells developing simultaneously. The measurements were obtained by a vertically-pointing Doppler radar located under the cloud at the position of the arrow. (From Battan and Theiss, 1966.)

those in Fig. 12.1. The downdraft is not yet apparent, possibly because it was displaced out of the observing plane of the radar.

The Byers–Braham thunderstorm cell is essentially a large shower, differing from the examples in Chapter 11 only in size and duration. While these are the most frequently occurring storms, it has been recognized for the past ten years that the most destructive thunderstorms are those in which the updraft and downdraft do not interfere with each other, but become organized to sustain a large, long-lasting convective circulation. The ambient winds are crucial in determining whether a thunderstorm will dissipate because of its downdraft or continue to exist.

Severe thunderstorms

Figure 12.3 is a radar example of two severe hailstorms which differ in structure and behavior from the thunderstorm cell. Because of the environmental wind, the updraft and downdraft are horizontally displaced from one another and mutually interact to sustain a strong, long-lived circulation. First analyzed by Browning and Ludlam (1962), this kind of storm circulation has become known as a "supercell".

Figure 12.4 is a schematic diagram showing horizontal sections of

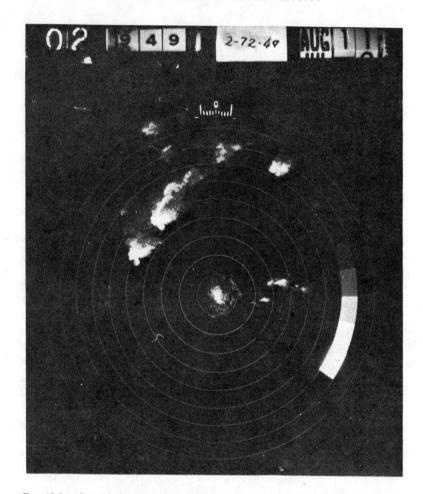

FIG. 12.3. Alberta hailstorms observed by radar at elevation angles of 2, 4, and 6 deg. Range rings at intervals of 10 miles; grey shades at 10-dB intervals. The two echoes to the northwest at about 50 miles range are similar to Browning's supercell.

a supercell echo at three altitudes. The key to this kind of storm is the ambient wind and its variation with height, indicated here by the vectors labeled *L*, *M*, and *H*, corresponding to the wind at low, medium, and high levels in the troposphere. The updraft enters at low levels and ascends in the region called by Browning the "vault".

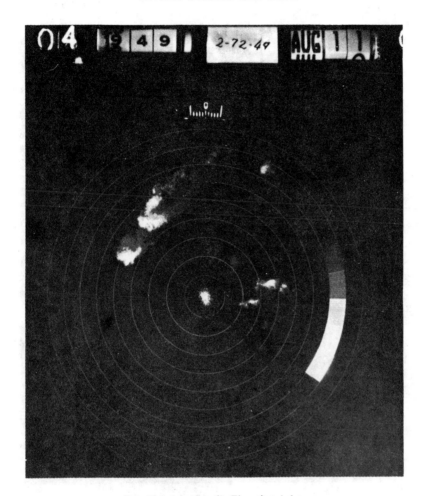

FIG. 12.3 (*continued*). Elevation 4 deg.

The updraft is so strong that precipitation is not able to grow to radar detectable size in the vault region. When precipitation does form at higher levels the wind shear prevents it from falling into the updraft at low levels and cutting off the circulation. This circulation is shown in more detail in Fig. 12.5.

Because of their size and destructiveness, supercell storms have

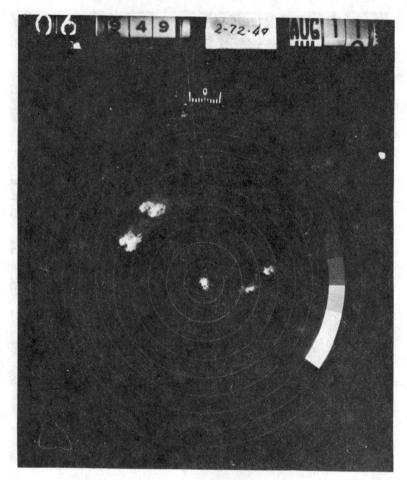

FIG. 12.3 (*continued*). Elevation 6 deg.

received much attention over the past few years. Yet they occur
rather infrequently, owing probably to the special wind pattern
required for their existence. A more frequently occurring storm,
which can also be large and severe, is the "multicell" storm (Fig.
12.6). Individual thunderstorm cells develop successively on the
right hand side of a large storm complex. Though each cell has a

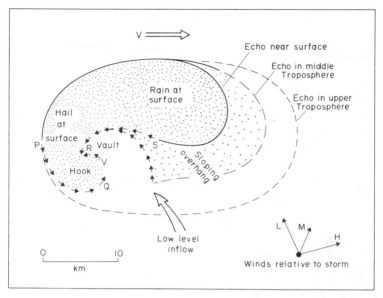

FIG. 12.4. Schematic diagram showing horizontal sections at three levels through the radar echo of a supercell storm. (From Browning, 1964.)

limited life cycle, the systematic development of new cells produces a long-lived storm. The distinction between supercell and multicell storms may not always be clear; some storms exhibit a supercell shape yet on close inspection are found to contain small-scale elements of short lifetime.

Hail growth

Hailstones are formed when either graupel particles or large frozen raindrops grow by accreting supercooled cloud droplets. Thunderstorms contain both graupel and large drops, and it is not known which serves most frequently as the hail "embryo", although photographic evidence generally points to graupel. An important aspect of hail growth is the latent heat of fusion released when the accreted water freezes. Owing to this heating, the temperature of a growing hailstone is several degrees warmer than its cloud

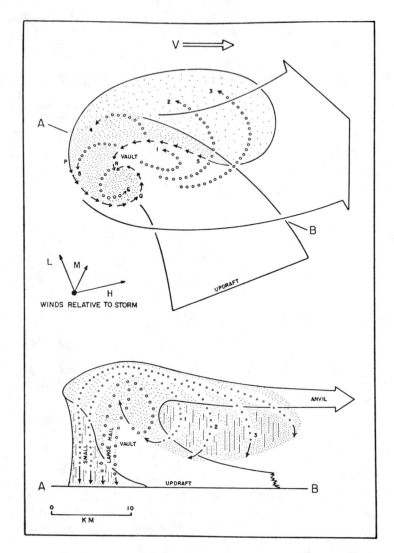

FIG. 12.5. Horizontal and vertical sections of airflow and precipitation trajectories in a supercell. (From Browning, 1964.)

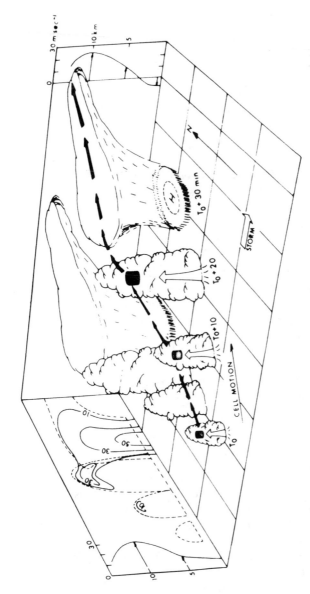

FIG. 12.6. Schematic view of a multicell storm. At the initial time the storm consists of four cells at different stages of development. The development of the youngest (southernmost) cell at successive times is indicated. The heavy dashed arrow is the trajectory of a parcel in the growing cell. A vertical section of the radar echo at the initial time is shown, as well as an indication of the wind profile. (From Chisholm and Renick, 1972.)

environment. In the theory of hail development, the temperature is determined by assuming a balance condition for the hailstone heating rate.

The heating rate due to the accretion of supercooled liquid droplets is given by

$$\frac{dQ_L}{dt} = \pi R^2 E M u(R)[L_f - c(T_s - T)], \qquad (12.1)$$

where R is the radius of the stone and $u(R)$ its fall speed, L_f is the latent heat of fusion, c the specific heat of water, M the cloud liquid water content and E the effective collection efficiency, and T_s and T the temperatures of the hailstone and the ambient cloud.

The heat gained by sublimation is

$$\frac{dQ_v}{dt} = 4\pi R D(\rho_v - \rho_{vR})L_s a, \qquad (12.2)$$

where ρ_v and ρ_{vR} denote the ambient vapor density and that at the surface of the stone, L_s is the latent heat of sublimation, and a is a ventilation factor depending on hailstone size.

The rate at which heat is lost to the air by conduction is

$$\frac{dQ_s}{dt} = 4\pi R K(T_s - T)b, \qquad (12.3)$$

where K is the heat conduction coefficient of air and b is a ventilation factor.

Equilibrium exists when

$$\frac{dQ_L}{dt} + \frac{dQ_v}{dt} = \frac{dQ_s}{dt}, \qquad (12.4)$$

which may be used to solve for the hailstone equilibrium temperature as a function of size, for given cloud conditions.

The rate of hailstone growth may be determined to a good approximation by adding the separate rates of growth by accretion and by sublimation. Accretion is usually dominant, and becomes more so as the stone grows. If it remains in the supercooled cloud long enough, the stone reaches the size for which the equilibrium temperature is 0°C, because of insufficient heat transfer to the surrounding air. (Typically this might occur for a diameter in the order of 1 cm.)

When the surface of the hailstone is at a subfreezing temperature, the collected water droplets freeze quickly and the surface remains essentially dry. When the surface is at 0°C, however, the collected water does not freeze immediately and the surface is wet. Although

1 cm

FIG. 12.7. Cross section of a large hailstone, showing the characteristic layered or "onionskin" structure. (Courtesy B. L. Barge, Alberta Research Council, Edmonton, Can.)

some water may be shed by the warm stone, much can remain to be incorporated into the stone, forming what is called "spongy ice". It has been deduced that the liquid fractions of large hailstones may amount to 20% or more. The entrapped liquid can later freeze if the stone enters colder or less dense cloud, where the heat transfer will suffice to chill the stone below 0°C. During its lifetime, a stone may undergo alternate wet and dry growth as it passes through a cloud of varying temperature and liquid water content, thus developing the layered structure that is often observed (Fig. 12.7).

Problems

1. A hailstone of initial radius 1 mm begins to fall from a height of 5 km above cloud base, where the ambient temperature is 250°K. It grows by accretion of cloud water under conditions such that its surface temperature is constant at 0°C. The air has a temperature lapse rate of 0.6°C/100 m within the cloud. Assume zero updraft velocity; assume also that during growth a balance always exists between the rate of heat gained by the freezing of accreted water and the rate of heat lost by conduction to the air. Neglect sublimation effects and the heat capacity of the collected water. Show that the hailstone grows to a radius of approximately 3.5 mm after falling 3 km. Appropriate values of the constants, in c.g.s. units, are as follows:

$$K = 2.3 \times 10^3$$
$$L_f = 3.3 \times 10^9$$
$$u(R) = kR^{1/2},$$

where $k = 2000$,

$$\text{ventilation factor } b = 0.3(Re)^{1/2},$$

where Re is the Reynolds number of the flow around the hailstone (Byers, 1965, p. 169).

2. A hailstone at 700 mb is falling through a cloud of supercooled droplets having a liquid water content of 4 g/m³ and a temperature of −15°C.

   Determine the dependence of hailstone temperature on size, assuming equilibrium growth conditions, over the temperature range from −15°C to 0°C. Take into account heat transfer by conduction, sublimation, and accretion. Use the same values for the physical constants as in Problem 1, and assume that the ventilation factors a and b are equal.

3. When falling through thin cloud or cloud-free air a hailstone is likely to be cooler than its environment because of sublimation from its surface and also because of its heat capacity, which leads to a time lag in its response to environmental temperature changes. Derive an expression for the thermal time-constant of a hailstone assuming an isothermal temperature structure within the stone at any time, and assuming heat transfer by conduction only. That is, in the hailstone heat transfer equation neglect all but the conduction term. Evaluate the time constant for a hailstone of 1 cm diameter, taking the specific heat of the stone to be 0.5 cal g^{-1} deg^{-1} and assuming a ventilation factor of 25.

CHAPTER 13

WEATHER MODIFICATION

WITH few exceptions, weather modification refers to the artificial modification of clouds. Experiments in cloud modification consist of seeding the cloud with some material, and are usually undertaken with one of the following goals in mind:

(1) to stimulate precipitation;
(2) to dissipate cloud or fog;
(3) to suppress hail.

Varying degrees of success have been reported for each type of experiment. There are theoretical bases for some forms of cloud modification; however, it is often difficult to obtain convincing evidence that an experiment was successful because of the large variability that characterizes clouds. Moreover, cloud seeding experiments are often conducted in a haphazard way, without the proper scientific controls to allow a sessment of results. Difficulties are further compounded by extravagant claims of success by some cloud-seeding practitioners. Nevertheless, evidence is emerging from a few carefully regulated experiments that, under the appropriate conditions, modest changes in cloud structure and precipitation can be effected by seeding. This chapter outlines some of the relevant physical principles.

Stimulation of rain and snow

In principle, precipitation may be encouraged by exploiting one of the instabilities in a cloud system. Only in this way is it reasonable to expect a large result from a relatively small expenditure of effort or material—or in other words to have an efficient modification technique. Before they produce precipitation, all clouds are colloidally unstable: the droplets have the potential of

194

being swept out by precipitation; all that is lacking are relatively few large drops which can grow by accretion of the droplets. In the natural rain process in warm clouds about one droplet in 10^6 grows to become a raindrop. This is about one droplet in each 5 liters or 200 per cubic meter. One approach to stimulating rain in warm clouds is therefore to introduce approximately this concentration of large droplets into the ascending air above cloud base. These drops stand some chance of ascending and sweeping out cloud droplets throughout their upward and downward trajectories. They must be large enough initially to be in a favored position for growth, but not so large that they fall out before spending much time in the cloud. Also, more water is required for seeding at the appropriate concentration if large drops are used. A suitable compromise is probably droplets in the radius interval 20–30 μm. Experiments in water-seeding have been attempted by spraying water from airplanes flying at cloud base, relying on diffusion to disperse the drops throughout the updraft region. These experiments have occasionally suggested modest precipitation enhancement, but have usually been inconclusive.

Alternatively, salt particles have been injected around cloud base to provide centers on which cloud droplets can form. Generally relatively large particles are preferred, since they will give solution droplets with relatively large critical radii. From (5.7) it follows that the critical radius r^* increases with salt particle radius according to $r^* \propto r^{3/2}$. The natural background of large salt particles (probably sea salt) is about 0.1 per liter. Therefore salt seeding should be designed to equal this amount at least, and to exceed it by perhaps an order of magnitude. In principle salt seeding is more efficient than water seeding, for a smaller mass of seeding material is needed to produce the same number of collector drops. However, there are practical problems connected with using salt, such as the clumping of particles in humid conditions and the corrosion of equipment, which do not exist for water. There are also theoretical uncertainties about the time required for salt particles of various sizes to grow to their critical droplet radii. Only rather recently have there been any encouraging results from salt seeding.

In subfreezing clouds there is thermodynamic phase instability associated with the presence of supercooled droplets. Once ice crystals form they will tend to grow rapidly by diffusion at the expense of these droplets. About one ice crystal per liter in the upper parts of a supercooled cloud is enough to lead to precipitation development. For temperatures of about $-20°C$ and colder this number will form due to natural nuclei. Introducing ice crystals artificially into warmer clouds, in concentrations of about 1 per liter, would be expected to promote precipitation. This technique would be relatively ineffective at temperatures warmer than about $-5°C$ because the ice crystal process is slow (see Fig. 8.2, p. 125).

The principal seeding agents used to form ice crystals are dry ice (solid CO_2) and silver iodide (AgI). The equilibrium temperature of subliming CO_2 is $-78°C$, considerably colder than even the homogeneous freezing temperature of water. Injected into a super-cooled cloud, pebble-size pieces of dry ice descend and freeze a large fraction of the droplets in their path. Laboratory experiments have shown that AgI particles can be introduced to clouds in the form of smoke produced by burning certain compounds of silver. Though their effectiveness is impaired by exposure to sunlight and possibly by wetting at warm temperatures, the AgI particles are known from observation to be an efficient source of ice crystals, at least in some conditions.

The possible effects of artificially produced ice crystals on a cloud depend on whether it is of cumuliform or stratiform type. In the special case of a stratiform cloud whose top is supercooled but does not extend beyond about the $-15°C$ level, the natural precipitation process may proceed very slowly owing to the scarcity of ice nuclei at relatively warm temperatures. In such clouds the introduction of ice crystals near cloud top by seeding with AgI or dry ice may cause precipitation that would not otherwise occur. The amount of precipitation would be small because of the low liquid water contents of cold stratiform clouds. If the cloud is persistent, however, due to large-scale meteorological factors, it would be possible by repeated seeding to build up a significant accumulation of precipitation. In stratiform clouds with tops colder than $-15°C$ seeding would not be required to initiate precipitation, but in some cases might be used to alter slightly the location where rain or snow falls—for example on a watershed instead of a few miles downwind.

For relatively small and short-lived cumuliform clouds the effect of seeding would be to initiate freezing slightly earlier and at a lower altitude than where it would occur naturally. It is not obvious that this would have much subsequent effect on cloud development except in very special circumstances. At most, seeding might cause a small amount of rain or snow to fall from a cloud that would otherwise fail to precipitate, or cause the rain to fall a little sooner than it would without seeding. Large and long-lasting cumulonimbus clouds produce precipitation naturally; seeding with ice nuclei would not be expected to affect the amount of precipitation produced.

Associated with freezing is the release of latent heat of fusion, representing what may be termed the latent instability inherent in supercooled clouds. This instability can be triggered by seeding with freezing nuclei or dry ice. It will contribute to the buoyancy of the updraft and may be of crucial importance in some clouds. Under some conditions cumulus clouds are limited in vertical development because of trapping by a thin inversion layer aloft. The extra latent heat liberated by seeding might be just sufficient, in a small class of such cases, to enable the cloud to penetrate the inversion and extend much further.

Cloud dissipation

Low-cloud overcasts and fogs pose hazards around airports. The concept of dissipating clouds by seeding is much like that of precipitation enhancement. Large particles or ice nuclei are introduced to sweep out the cloud droplets, thus clearing an area temporarily. Silver iodide and dry ice have been used with some success for clearing supercooled fogs and overcasts. Warm fogs have proved more difficult to affect. Experiments using salt and sprays have been attempted since as early as 1938 (Houghton and Radford), but there are no operational systems employing this approach. Ice fog, a winter phenomenon in some northern localities, also remains an unsolved problem.

Hail suppression

Two arguments have been advanced for alleviating hail by cloud seeding with ice nuclei. The first involves freezing essentially all of the supercooled droplets in the upper parts of a potentially hail-producing cumulonimbus. This in effect kills the accretional growth process, eliminating the possibility of large hail formation. Although the nucleating efficiency of AgI is presumably high, with estimates typically about 10^{14} nuclei per gram of AgI at $-20°C$, the amount of material required to glaciate a cloud is excessive and much beyond the capability of any seeding system in current use.

The second argument is more modest in the requirement of seeding material and involves adding ice nuclei only to the limited cloud region where hail is thought to have its maximum growth rate. Soviet scientists assume that this is the region in the upper part of the cloud where maxima in radar reflectivity are occasionally observed. They seed this region with AgI-charged artillery shells and report spectacular success in eliminating hail. The reason for the apparent effectiveness of this technique is not entirely clear.

A variation of the Soviet approach is to add the ice nuclei in a region lower down which is presumed to be the main updraft area. This region contains the natural ice nuclei or precipitation particles that are hail embryos. It is argued that by introducing artificial nuclei it may be possible to cause enough competition for the available water supply to make it unlikely that any hailstone grows to a large size. Therefore, what this approach intends to cause is a large number of small stones instead of a few large ones. The small stones stand a chance of melting completely before reaching ground, or at least of causing less damage than the large ones.

Problems

1. One of the proposals put forward for suppressing hail is to "seed" the storm with ice nuclei in the updraft region where the natural ice crystals that develop into hailstones originate. This idea assumes that the artificial hail "embryos" will grow in the ascending air as graupel particles along with the natural embryos, leading to a high concentration of small hailstones in the cloud region of high liquid water content, where accretional growth would ordinarily be most rapid. These small hailstones compete for the available liquid water, and if their concentration is high enough none will be able to grow to the size of large, damaging hail. The critical question in this scheme is how many artificial embryos to introduce.

   One way to approach the problem is to compare at a given altitude the rate at which liquid water is being made available by condensation in the updraft with the rate at which this water is used up by accretion. When the concentration of small hailstones exceeds some critical value, the depletion rate will exceed the production rate. This may be looked upon as the *minimum* hailstone concentration for hail suppression to be effective.

Set up the equations needed to analyze the problem from this point of view and estimate the minimum required hailstone concentration under the following assumptions:

(a) all hailstones are the same size, spherical, with diameter of 2 mm and terminal fall speed of 8 m/sec,
(b) altitude about 18,000 ft (500 mb pressure),
(c) updraft speed 25 m/sec,
(d) growth by sublimation negligible,
(e) cloud liquid water content 6 g/m³,
(f) collection efficiency of unity.

2. The supercooled portion of a convective cloud is rapidly converted to ice by seeding with a material that promotes freezing. At the 600 mb level the cloud temperature is initially −20°C and the liquid water content is 3 g/kg. Calculate the total increase in temperature at this level as a result of complete glaciation of the cloud liquid.

Saturation mixing ratios over ice and water at 600 mb are given for various temperatures in the following table:

T (°C)	w_s (g/kg)	w_i (g/kg)
− 23	1.002	0.800
− 22	1.094	0.883
− 21	1.194	0.973
− 20	1.303	1.072
− 19	1.420	1.179
− 18	1.546	1.296
− 17	1.682	1.425
− 16	1.824	1.565

CHAPTER 14

NUMERICAL CLOUD MODELS

MOST clouds form in association with the expansion and cooling of ascending air. The only important exception is fog, certain varieties of which form by radiational cooling of the air near the ground or by the mixing of airmasses with different temperatures. In the case of stratiform clouds the ascent is controlled by large-scale processes such as uplifting along a tilted frontal surface or slow ascent in a field of cyclone-scale convergence. In cumuliform clouds the upward motion is due to convection in unstable air.

Large-scale air motions are understood in terms of the equations of dynamic meteorology. Convection is not so well understood because of the increased importance of small-scale turbulent effects. When cloud is present, the air motions become even more difficult to describe mathematically because of the latent heat accompanying phase transitions and the drag exerted on the air by the condensed material. Nevertheless, theoretical models of clouds—especially convective clouds—have been developed over the past decade which approximate the behavior of natural clouds in some respects. As an important area of current meteorological research, these models are being steadily refined and improved.

Three kinds of processes must be accounted for in the models: dynamic, thermodynamic, and cloud physical. Most of these are reasonably well understood. The problem is then to express the processes in terms of a system of differential equations that can be solved. The equations that result can only be solved numerically, and this is one of the difficulties in theoretical cloud modeling. Considerable effort is required to insure that solutions do not depend on the particular numerical approximations used. Another important difficulty is that the dynamic process of mixing between the cloud and outside air is not adequately understood. Mixing effects are included in some of the models, but often by rather arbitrary or empirical approximations.

Research has usually been directed to the simulation of single, isolated cumulus clouds. This makes it possible to treat the cloud as independent of its surroundings, though in fact natural clouds may often influence each other. Even with this approximation the volume in space which needs to be considered (the domain of the calculations) is in the order of 5×10^{10} m^3 or larger, and it may be desired to study the cloud as it develops over some tens of minutes. By numerical process the quantities calculated (e.g. temperature, air velocity, cloud water content) are determined at grid points in time and space. The computer time required for a simulation increases rapidly with the number of grid points. In order to keep the time reasonably short it is therefore necessary that relatively few grid points be used or that other simplifications be made. Inevitably this introduces uncertainty. For example, although much of the energy in turbulent cloud motions is contained in eddies or irregularities smaller than 100 m, the grid spacing in most cloud models is larger than 100 m. The small eddies are responsible for mixing but cannot be resolved in the usual grid; a continuing concern has been how to approximate the effects of turbulence without actually solving for the eddy structure. Another way to reduce computation time is to consider a cloud in one or two dimensions instead of three. Again, however, this leads to uncertainty because of mounting evidence that models with fewer dimensions than three cannot properly account for the dynamic effects.

The purpose of this chapter is to outline the fundamentals of cloud modeling and illustrate a few results. Numerical and computational details, though an important part of the subject, are not included. For this information the reader must turn to the references. For convenience, the models are classified according to the number of spatial dimensions and the complexity of cloud physical processes that are included.

One-dimensional models with simplified microphysics

In cumulus clouds there is a strong interaction between dynamics and microphysics. The updraft controls the development of precipitation, but in growing and falling out the precipitation interferes with the updraft. These interactions were first simulated in one-dimensional cloud models, in which both the dynamics and microphysics are simplified. Cloud quantities are assumed to vary only in the vertical, and the ambient conditions are assumed not to change with time. Freezing is neglected. Water may be present in the form of vapor, cloud water, and rainwater. The sizes of cloud drops and raindrops are not specified; however, cloud drops move directly with the air and raindrops fall relative to the air with a velocity that depends on the amount of rainwater per unit volume. Details of the condensation process are neglected: any excess in vapor over the equilibrium value is assumed to condense out in the form of cloud water. Similarly, any deficit below the equilibrium value is supplied by evaporation of the available condensed water. Coalescence is neglected: cloud water is assumed to convert spontaneously to rainwater once a specified threshold in cloud water content is exceeded.

Representative of this kind of model is the one developed by Srivastava (1967). With the various approximations, the model reduces to a system of five equations in five unknowns, the vertical velocity U, temperature T, vapor mixing ratio X, cloud water mixing ratio W, and rainwater mixing ratio R. Each of these quantities is a function of height and time. The governing equations, as restated by Harris (1969), are as follows:

Velocity:
$$\frac{\partial U}{\partial t} = -U \frac{\partial U}{\partial z} + gB - g(W + R), \qquad (14.1)$$

where $B = (T - T')/T'$, with $T(z,t)$ the temperature in the cloud and $T'(z)$ the ambient temperature. The third term on the right represents the reduction in buoyancy due to the weight of condensed water; without this term (14.1) would be simply the equation of motion for a buoyant parcel.

Water vapor:
$$\frac{\partial X}{\partial t} = -U \frac{\partial X}{\partial z} + E, \qquad (14.2)$$

where E is an evaporation term that can be positive or negative.

Cloud water:
$$\frac{\partial W}{\partial t} = -U \frac{\partial W}{\partial z} - E_1 - P, \qquad (14.3)$$

where E_1 is a term that describes cloud evaporation or condensation, and P describes the production of rain from cloud by spontaneous coalescence (called autoconversion) and by accretion.

Rainwater:
$$\frac{\partial R}{\partial t} = -U\frac{\partial R}{\partial z} - \frac{1}{\rho}\frac{\partial}{\partial z}(\rho R V) - E_2 + P, \qquad (14.4)$$

where E_2 is a term that describes the rate of evaporation of rain. In the second term ρ is air density and V is the effective fall velocity of the rain; this term represents the vertical divergence of rainwater.

Temperature:
$$\frac{\partial T}{\partial t} = -U\frac{\partial T}{\partial z} - U\Gamma - \frac{L}{c_p}E, \qquad (14.5)$$

where $E = E_1 + E_2$ is the evaporation factor also appearing in (14.2), and Γ is the dry adiabatic lapse rate.

The factors E_1, E_2, V, and P were expressed by Kessler (1959, 1969) as functions of the other dependent variables in the system on the basis of empirical arguments about drop-size distributions, fall velocities, and the conversion of cloud to rain. Kessler's expressions vastly simplify calculations, making it unnecessary to account specifically for drops of different sizes. These parameterizations have been employed directly or in modified form in most of the simpler cloud models, including that of Srivastava.

Equations (14.1)–(14.5) were solved by Srivastava for various combinations of initial conditions and assumptions. Figure 14.1 shows results from a case with the following initial conditions: the initial profiles of updraft and cloud water are the steady state solutions of (14.1) and (14.3) with R set equal to zero; no rain is present initially; U and W are initially zero at cloud base; U is zero at the surface and remains so. The ambient temperature sounding is representative of convective rain conditions in warm climates, with a cloud base temperature of 4°C and a lapse rate of 6.8°C/km. At all altitudes except the very top of the cloud, the updraft is a maximum initially, when no rain is present. During the first 10 min rain develops at the expense of cloud water through the entire cloud, with the maximum rain production in the upper part of the cloud, where the cloud water content was initially a maximum. At 15 and 25 min the rain has descended, creating a low-level downdraft.

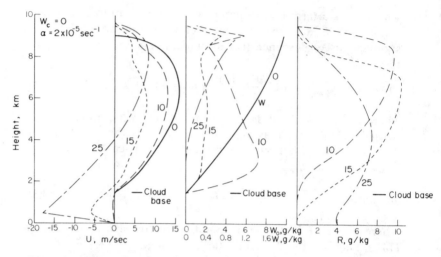

FIG. 14.1. Results of calculations for a one-dimensional cloud model, showing vertical profiles of updraft (left), cloud water mixing ratio (center), and rainwater mixing ratio (right) at the indicated times in minutes. Note that different scales are used for the cloud water content initially (W_0) and at subsequent times (W). (From Srivastava, 1967.)

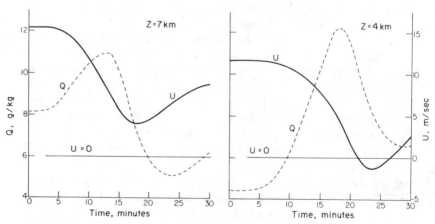

FIG. 14.2 Variation in time of updraft velocity U and total water content Q (cloud plus rain) at heights of 7 km (left) and 4 km (right). (From Srivastava, 1967.)

The interaction between water content and vertical velocity is illustrated in Fig. 14.2. At both altitudes depicted the updraft velocity weakens as the water content increases. After the water content reaches a peak, the updraft is allowed to increase, but this in turn increases the rate of production of water. Thus the feedback works in such a way that the water content lags behind the updraft at a given level.

One-dimensional models with more complex microphysics

Although one-dimensional models can account only crudely for dynamic processes, it is possible in such models to include detailed microphysical processes and still require less computer time than the multi-dimensional models. For example, Ogura and Takahashi (1971) developed a model for a thunderstorm cell including the ice phase. Thus, in addition to the effects considered by Srivastava, they included freezing, sublimation, and melting in their model. Again, however, particles of different size were not specifically accounted for; a parameterized approach similar to Kessler's was used to approximate the microphysics. Whereas Srivastava did not attempt to simulate entrainment-mixing, Ogura and Takahashi included an entrainment term which is inversely proportional to the horizontal extent of the buoyant element.

Figure 14.3 shows the results for a typical case. The model is initiated by a small velocity perturbation near cloud base. Since the atmosphere is conditionally unstable, the perturbation amplifies and the updraft intensifies for the first 35 min. The condensed water reduces the buoyancy, leading to the formation of downdrafts after about an hour. The model satisfactorily approximates some of the observed characteristics of thunderstorm cells—in particular the general form of the life cycle, including the developing, mature, and dissipating stages, and also the magnitudes of quantities such as updraft speed and water content. Ogura and Takahashi found that the most important parameter in determining the life cycle of the cloud cell is the rate of conversion from cloud drops to raindrops. Freezing was also important, tending to increase both the maximum updraft and the maximum precipitation rate at the surface. Evaporation and sublimation were of little importance.

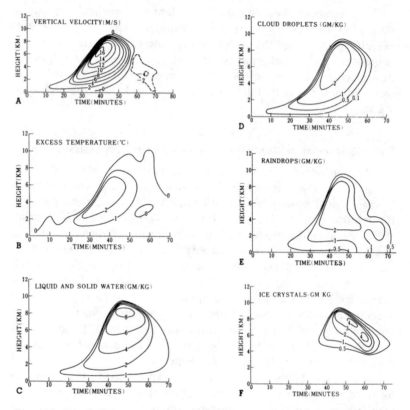

FIG. 14.3. Results from a one-dimensional cloud model including ice-phase microphysics: time-height cross-sections of the indicated quantities. (From Ogura and
Takahashi, 1971.)

Danielsen *et al.* (1972) have developed a cloud model that goes
much further into microphysical details. The water and ice phases
are both included and the size distributions of water drops and ice
particles are explicitly taken into account. Only the initial droplet
distribution at cloud base must be specified; development then
proceeds by condensation, stochastic coalescence, and freezing.
The distribution of water drops is defined by 31 categories of size,
arranged logarithmically from 2.5 μm to 2.5 mm in radius. Nine
additional categories extend the ice particles to a radius of 2 cm.

With the distribution thus described it is not necessary to employ parameterizations for autoconversion, condensation, and evaporation, and the important cloud-to-rain conversion can be determined more accurately than in the simpler, parameterized models. Entrainment is simulated by a method similar to that of Ogura and Takahashi. Breakup of large drops is modeled by forcing distributions toward the Marshall–Palmer negative-exponential form for large drop sizes. An empirical method is also used in modeling freezing, such that freezing becomes possible as the temperature drops below $-7°C$, with large drops being preferentially frozen first, and all drops are frozen for $T < -45°C$.

Figure 14.4 shows some of the results of Danielsen et al., for a case in which the simulated cloud corresponded rather closely to an actual observed hailstorm. It was found that hail growth in the model depends on the updraft speeds, initial droplet distribution, surface mixing ratio, and height above cloud base at which drops begin to freeze. Hail was favored in cases with a broad initial drop spectrum in which the maximum updraft velocity is between 15 and 30 m/sec. Maximum hail growth occurs in the mixed-phase region of the cloud, which is located beneath the updraft maximum.

Two-dimensional cloud models

In long-lasting convective clouds the downdraft is horizontally displaced from the updraft because of the effect of wind shear on precipitation trajectories. One-dimensional models cannot include vertical shear of the ambient wind, and thus cannot account for an important characteristic of large storm clouds. Accordingly, a number of investigators have devised models with two spatial dimensions in which the effects of ambient shear on cloud development can be assessed. In most of these the rate of conversion from cloud to rain is assumed and microphysical processes are parameterized. An exception is the model of Takeda (1971) in which the microphysical processes are given a more realistic treatment, although the ice phase is neglected.

In two dimensions, quantities are computed as functions of time in a vertical plane instead of simply along a vertical line. Also both the horizontal and vertical components of velocity in the plane must

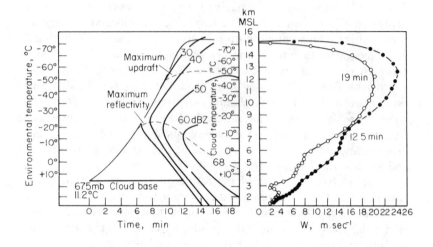

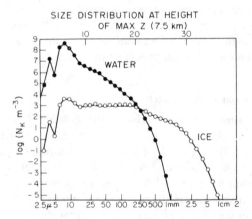

FIG. 14.4. Results from a model including extensive microphysical details: above, time-height pattern of radar reflectivity and updraft profiles; below, size distribution of water drops and ice particles at level of maximum reflectivity. (From Danielsen *et al.*, 1972.)

be computed. It is assumed that there is no flow into or out of the plane and that none of the quantities varies in the direction normal to the plane. Since the flow is confined to two dimensions, it is possible to express both velocity components in terms of a single stream function, which simplifies the calculations. An economy may also be realized by combining the two equations of motion into a single vorticity equation. Additional equations are needed, of course, to describe the thermodynamic and cloud physical processes.

In Takeda's model the size distribution of water drops is described in terms of seven discrete radii. Cloud and rainwater contents, rainfall intensity, and radar reflectivity are computed directly from the size distributions. The model includes condensation, evaporation, stochastic coalescence, spontaneous breakup, and the fall velocity of drops relative to the air. Dynamically, diffusion is modeled by the inclusion of an eddy diffusion coefficient in the governing equations.

Convection is started by an initial disturbance of temperature, the extent and intensity of which are specified. Although the main interest was in the effects of ambient wind, Takeda also experimented with different values of the ambient stability and of the initial disturbance. He found that the clouds could be grouped into three types according to the way the cloud develops: type 1, in which new clouds develop around a dissipating cloud; type 2, in which the cloud is short-lived; and type 3, in which the cloud is long-lasting.

Type 1 clouds develop if the atmosphere is sufficiently unstable and the ambient wind shear is weak. A downdraft forms in the initial cloud because of precipitation loading and cooling by evaporation. The current of cold air from the downdraft spreads outward near the ground, causes lifting of the surrounding air, and stimulates the development of a new convective cloud.

If the wind is strongly sheared in the vertical, with the velocity monotonically increasing with height, the initial cloud is inclined in the same sense as the shear. The precipitation-induced downdraft forms on the downshear side of the cloud, and the convective upcurrent is tilted by the vertical wind shear. This arrangement intensifies the vertical shear on the downshear side of the cloud. Initiated by the cold air outflow, new upcurrents are damped by the intensified shear before they can develop a convective cloud circulation. The cloud is therefore short-lived.

When the wind profile consists of a jet or extremal value at lower levels, with layers above and below the jet having shear vectors in opposite directions, the patterns of rainwater content and updraft velocity are both inclined in the direction of the lower level shear. The rainwater is displaced from the updraft in the direction of the jet, leading to the formation of a downdraft. The dome of cold air originating from the downdraft creates a new upcurrent in the vicinity of the original updraft. In this situation the updraft and downdraft form an organized circulation that can last a long time. The convective cloud attains a steady state and is termed long-lasting. If the height of the jet is too low or too high it has little effect on the convection and a type-2 cloud develops.

Figure 14.5 illustrates the rainwater content and streamlines for a type-3 cloud in an environment with a low-level jet at a height of 2.5 km. By 30 min the precipitation-induced downdraft has been formed to the left of the upcurrent. This induces low-level upward motion to the left of the cloud, leading to the formation of a small raincloud at 45 min. The small cloud does not develop further, and at 60 min the structure of the parent cloud is similar to that at 30 min.

Takeda was among the first to include radar reflectivity in a cloud model. This is an important quantity to compute, because it is often the only means of comparing the structure of real clouds with model predictions. Takeda's calculated reflectivities agreed well in magnitude and time and space structure with those observed in large convective clouds.

Three-dimensional models

Although more realistic than models in only one dimension, two-dimensional models have certain shortcomings. Confining the flow to a plane means that the circulation must be closed in the plane parallel to the wind. That is, the downdraft is forced to form in the same plane as the updraft. In the real atmosphere the wind field is more complex, and the circulations observed in large convective clouds are markedly three-dimensional, with low level inflow generally in a different plane from high level outflow. Strictly speaking, two-dimensional models could simulate only roll-type clouds, in which most variations are confined to the vertical plane normal to the roll axis. Apart from the questions of realism, there are

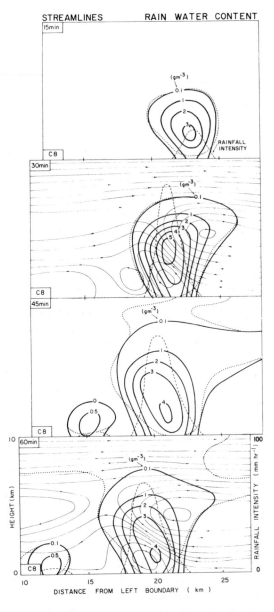

FIG. 14.5. Rainwater contents (g/m³) and streamlines in a simulated two-dimensional precipitating cloud growing in a sheared environment. The broken line indicates rainfall intensity at the 500 m level, and the dotted line shows the cloud boundary. (From Takeda, 1971.)

uncertainties whether the circulations that develop in two-dimensional models would even occur if motion in the third dimension was permitted.

The simplest three-dimensional models make the assumption of axial symmetry. Cylindrical coordinates are used, and all dependent variables are constant with respect to the angle measured outwards from the vertical axis. The circulation in three dimensions may then be represented by only a two-dimensional array of grid points. Murray and Koenig (1972) employed such a model to investigate the effects of parameterized microphysical processes on cumulus development. They found that evaporation was one of the more important processes affecting cloud behavior. As the cloud develops, lifting and evaporation at the top produce a cold cap, and this cold, dry air is brought down along the edge of the cloud where it mixes with cloudy air to produce further evaporation and cooling. This growing cap of cold, dense air eventually stops the upward development of the cloud. Evaporation of rain falling below cloud base establishes a downdraft and a relatively strong circulation cell. Away from the cloud axis the downdraft extends to higher levels than it does at the axis, cutting off the inflow to the lower part of the cloud but permitting some inflow at middle levels. After about 30 min all inflow is cut off and thereafter the cloud decays.

Axially symmetric models are not able to include ambient wind shear. They are suitable for simulating "single-cell" convective clouds that form in an unsheared environment, but cannot duplicate the larger, longer lasting convective clouds in which wind shear plays an essential role. A full, three-dimensional model including precipitation has not yet been developed, although Steiner (1973) has made a significant step in that direction with a non-precipitating cloud model in three dimensions.

Figure 14.6 shows the patterns of vertical velocity and cloud water content for a cloud developing in a sheared environment. The cloud is initiated by a perturbation in the humidity field, which defines a region of excess virtual temperature. The ambient sounding was representative of maritime conditions in which cumulus clouds producing little or no rain form. The strongest downdraft develops near the top and on the right-hand (downshear) side of the cloud. The close proximity of updraft and downdraft maxima

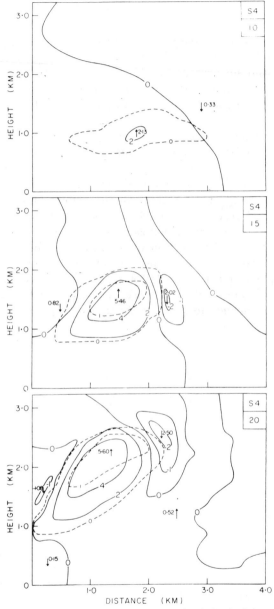

FIG. 14.6. Vertical cross-sections at 10, 15, and 20 min of cloud mixing ratio (dashed lines, g/kg) and vertical velocity (solid lines, m/sec) in a three-dimensional cloud model. Arrows indicate vertical velocity extremes. The cross-sections are through the center of the cloud in the plane of the ambient wind. (From Steiner, 1973.)

creates a region of intense mixing between cloud and environment at the upper level, downshear side. Comparing cloud development in sheared and unsheared flow, Steiner found that shear dampens convective development even though in the sheared case there is a transfer of kinetic energy from the mean flow to the convective motion.

To assess the importance of including the third dimension, Steiner made calculations with the same ambient conditions as in the case of Fig. 14.6 but with the domain width reduced to only one grid length in the direction perpendicular to the shear. This, in effect, contracts the three-dimensional model to two dimensions. This change made a significant difference in cloud behavior. In two dimensions the magnitudes of vertical velocities were reduced and the sizes, and in some cases even the direction, of energy conversions were different. While it is possible with simpler models to investigate precipitation development and, to a limited extent, interactions between cloud dynamics and microphysics, it becomes clear that three-dimensional models are needed to simulate the dynamics properly.

REFERENCES

AUFM KAMPE, H. J., and H. K. WEICKMANN (1957) *Physics of Clouds.* Meteor. Monograph, Vol. 3, No. 18. American Meteorological Society, Boston, pp. 182–225.

AUSTIN, P. M., and R. A. HOUZE (1972) Analysis of the structure of precipitation patterns in New England. *J. Appl. Meteor.* **11**, 926–935.

AUSTIN, P. M., and M. J. KRAUS (1968) Snowflake aggregation—a numerical model. *Proc. Intl. Conf. on Cloud Phys.*, Toronto, pp. 300–304.

BATTAN, L. J. (1973) *Radar Observation of the Atmosphere.* University of Chicago Press, 324 pp.

BATTAN, L. J., and J. B. THEISS (1966) Observations of vertical motions and particle sizes in a thunderstorm. *J. Atmos. Sci.* **23**, 78–87.

BEARD, K. V. (1976) Terminal velocity and shape of cloud and precipitation drops aloft. *J. Atmos. Sci.* **33**, 851–864.

BIGG, E. K. and J. GUITRONICH (1967) Ice nucleating properties of meteoric material. *J. Atmos. Sci.* **24**, 46–49.

BOROVIKOV, A. M., A. KH. KHRGIAN and others (1961) *Cloud Physics.* U.S. Dept. of Commerce, Washington D.C., 392 pp. (Tr. from Russian by Israel Program for Scientific Translations.)

BRAHAM, R. R. (1952) The water and energy budgets of the thunderstorm and their relation to thunderstorm development. *J. Meteor.* **9**, 227–242.

BRAHAM, R. R. (1963) Phloroglucinol seeding of undercooled clouds. *J. Atmos. Sci.* **20**, 563–568.

BRAHAM, R. R. (1968) Meteorological bases for precipitation development. *Bull. Amer. Meteor. Soc.* **49**, 343–353.

BRAZIER-SMITH, P. R., S. G. JENNINGS, and J. LATHAM (1972) The interaction of falling water drops: coalescence. *Proc. Roy. Soc. Lond.* **A326**, 393–408.

BRAZIER-SMITH, P. R., S. G. JENNINGS, and J. LATHAM (1973) Raindrop interactions and rainfall rates within clouds. *Quart. J. Roy. Meteor. Soc.* **99**, 260–272.

BROCK, J. R. (1972) Condensational growth of atmospheric aerosols. *Aerosols and Atmospheric Chemistry* (G. M. Hidy, Ed.), Academic Press, New York, 149–153.

BROWNING, K. A. (1964) Airflow and precipitation trajectories within severe local storms which travel to the right of the winds. *J. Atmos. Sci.* **21**, 634–639.

BROWNING, K. A., T. W. HARROLD, A. J. WHYMAN, and J. G. D. BEIMERS (1968) Horizontal and vertical air motion and precipitation growth, within a shower. *Quart. J. Roy. Meteor. Soc.* **94**, 498–509.

BROWNING, K. A., and T. W. HARROLD (1969) Air motion and precipitation growth in a wave depression. *Quart. J. Roy. Meteor. Soc.* **95**, 288–309.

BROWNING, K. A., and F. H. LUDLAM (1962) Airflow in convective storms. *Quart. J. Roy. Meteor. Soc.* **88**, 117–135.

BYERS, H. R. (1965) *Elements of Cloud Physics.* University of Chicago Press, 191 pp.

BYERS, H. R., and R. R. BRAHAM (1949) *The Thunderstorm.* U.S. Dept. of Commerce, Washington, 287 pp.

CHISHOLM, A. J. (1973) *Alberta Hailstorms. Part I: Radar Case Studies and Airflow Models.* Meteor. Monograph, Vol. 14, No. 36. American Meteorological Society, Boston, pp. 1–36.

CHISHOLM, A. J., and J. H. RENICK (1972) Supercell and multicell Alberta hailstorms. Abstracts, Intl. Conf. on Cloud Physics, London, pp. 67–68.

COTTON, W. R., J. E. JIUSTO, and R. C. SRIVASTAVA (1975) Cloud physics and radar meteorology. *Rev. Geophys. and Space Phys.* **13**, 753–760.

DANIELSEN, E. F., R. BLECK, and D. A. MORRIS (1972) Hail growth by stochastic coalescence in a cumulus model. *J. Atmos. Sci.* **29**, 135–155.

DAVIS, M. H., and J. D. SARTOR (1967) Theoretical collision efficiencies for small cloud droplets in Stokes flow. *Nature* **215**, 1371–1372.

DOUGLAS, R. H., K. L. S. GUNN, and J. S. MARSHALL (1957) Pattern in the vertical of snow generation. *J. Meteor.* **14**, 95–114.

DOUGLAS, R. H., and J. S. MARSHALL (1954) The convection associated with the release of latent heat of sublimation. McGill University, Stormy Weather Group Rep. MW-19, 31 pp.

DRAKE, R. L. (1972a) A general mathematical survey of the coagulation equation. *Topics in Current Aerosol Research* (*Part* 2) (G. M. Hidy & J. R. Brock, Eds.), Pergamon Press, Oxford, 201–384.

DRAKE, R. L. (1972b) The scalar transport equation of coalescence theory: moments and kernels. *J. Atmos. Sci.* **29**, 537–547.

EAST, T. W. R. (1957) An inherent precipitation mechanism in cumulus clouds. *Quart. J. Roy. Meteor. Soc.* **83**, 61–76.

FITZGERALD, J. W. (1972) A study of the initial phase of cloud droplet growth by condensation and comparison between theory and observation. Ph.D. thesis, University of Chicago, 144 pp.

FLETCHER, N. H. (1962) *The Physics of Rainclouds.* Cambridge University Press, 386 pp.

FRENKEL, J. (1946) *Kinetic Theory of Liquids.* Dover Publications, New York, 488 pp.

FUKUTA, N. (1966) Experimental studies of organic ice nuclei. *J. Atmos. Sci.* **23**, 191–196.

FUKUTA, N. (1969) Experimental studies on the growth of small ice crystals. *J. Atmos. Sci.* **26**, 522–531.

FUKUTA, N., and L. A. WALTER (1970) Kinetics of hydrometeor growth from a vapor-spherical model. *J. Atmos. Sci.* **27**, 1160–1172.

GILLESPIE, D. T. (1972) The stochastic coalescence model for cloud droplet growth. *J. Atmos. Sci.* **29**, 1496–1510.

GILLESPIE, D. T. (1975) Three models for the coalescence growth of cloud drops. *J. Atmos. Sci.* **32**, 600–607.

GUNN, R., and G. D. KINZER (1949) The terminal velocity of fall for water drops in stagnant air. *J. Meteor.* **6**, 243–248.

GUNN, K. L. S., and J. S. MARSHALL (1958) The distribution with size of aggregate snowflakes. *J. Meteor.* **15**, 452–461.

HALLETT, J., and S. C. MOSSOP (1974) Production of secondary ice crystals during the riming process. *Nature* **249**, 26–28.

HARDY, K. R. (1963) The development of raindrop-size distribution and implications related to the physics of precipitation. *J. Atmos. Sci.* **20**, 299–312.

HARRIS, F. I. (1969) Further studies of the effects of precipitation on convection. McGill University, Stormy Weather Group Rep. MW-60, 25 pp.

HARROLD, T. W., and P. M. AUSTIN (1974) The structure of precipitation systems—a review. *J. Rech. Atmos.* **8**, 41–57.

HIDY, G. M., and J. R. BROCK (1970) *The Dynamics of Aerocolloidal Systems.* Pergamon Press, Oxford, 379 pp.

HIRTH, J. P., and G. M. POUND (1963) *Condensation and Evaporation, Nucleation and Growth Kinetics.* Pergamon Press, Oxford, 177 pp.

HOBBS, P. V., and W. M. KETCHAM (1969) Physics of ice. *Proc. Intl. Symp. on Physics of Ice,* Munich, 1968. Ed. N. Riehl, B. Bullemer, and H. Engelhardt. Plenum Press, New York, pp. 95–112.

HOCKING, L. M. (1959) The collision efficiency of small drops. *Quart. J. Roy. Meteor. Soc.* **85**, 44–50.

HOCKING, L. M., and P. R. JONAS (1970) The collision efficiency of small drops. *Quart. J. Roy. Soc.* **96**, 722–729.

HOUGHTON, H. G. (1950) A preliminary quantitative analysis of precipitation mechanisms. *J. Meteor.* **7**, 363–369.

HOUGHTON, H. G. (1968) On precipitation mechanisms and their artificial modification. *J. Appl. Meteor.* **7**, 851–859.

HOUGHTON, H. G., and W. H. RADFORD (1938) On the local dissipation of natural fog. Papers in Phys. Ocean. and Meteor., Mass. Inst. of Technology and Woods Hole Ocean. Inst. 6, No. 3, 63 pp.

HOWELL, W. E. (1949) The growth of cloud drops in uniformly cooled air. *J. Meteor.* **6**, 134–149.

JIUSTO, J. E. (1966) Aerosol and cloud physics measurements in Hawaii. *Tellus* **19**, 359–368.

JIUSTO, J. E., and H. K. WEICKMANN (1973) Types of snowfall. *Bull. Amer. Meteor. Soc.* **54**, 1148–1162.

JOSS, J., J. C. THAMS, and A. WALDVOGEL (1968) The variation of raindrop size distribution at Locarno. *Proc. Intl. Conf. on Cloud Phys.,* Toronto, pp. 369–373.

KESSLER, E. (1959) Kinematical relations between wind and precipitation distributions. *J. Meteor.* **16**, 630–637.

KESSLER, E. (1969) *On the Distribution and Continuity of Water Substance in Atmospheric Circulations.* Met. Monograph, Vol. 10, No. 32, American Meteorological Society, Boston, 84 pp.

KOMABAYASI, M., T. GONDA, and K. ISONO (1964) Life times of water drops before breaking and size distribution of fragment droplets. *J. Meteor. Soc. Japan.* **42**, 330–340.

KOVETZ, A. (1969) An analytical solution for the change of cloud and fog droplet spectra due to condensation. *J. Atmos. Sci.* **26**, 302–304.

KOVETZ, A., and B. OLUND (1969) The effect of coalescence and condensation on rain formation in a cloud of finite vertical extent. *J. Atmos. Sci.* **26**, 1060–1065.

KUHNS, I. E., and B. J. MASON (1968) The supercooling and freezing of small water droplets in air. *Proc. Roy. Soc.* **A302**, 437–452.

KUMAI, M. (1967) Fog modification on the Greeland Ice Cap. *Proc. First Natl. Conf. on Wea. Mod.,* Albany, N.Y., pp. 414–422.

LAMB, H., (1945) *Hydrodynamics.* Dover Publications, New York, 738 pp.

LANGER, G., J. ROSINSKI, and S. BERNSEN (1963) Organic crystals as icing nuclei. *J. Atmos. Sci.* **20**, 557–562.

LANGLEBEN, M. P. (1954) The terminal velocity of snowflakes. *Quart. J. Roy. Meteor. Soc.* **80**, 174–181.

LEIGHTON, H. G., and R. R. ROGERS (1974) Droplet growth by condensation and coalescence in a strong updraft. *J. Atmos. Sci.* **31**, 271–279.

LIST, R. J. (ed.) (1958) *Smithsonian Meteorological Tables.* Smithsonian Institution, Washington, 527 pp.

LONG, A. B., (1971) Validity of the finite-difference droplet collection equation. *J. Atmos. Sci.* **28**, 210–218.

MARSHALL, J. S. (1953) Precipitation trajectories and patterns. *J. Meteor.* **10**, 25–29.

MARSHALL, J. S., and W. MCK. PALMER (1948) The distribution of raindrops with size. *J. Meteor.* **5**, 165–166.

MASON, B. J. (1952) The production of rain and drizzle by coalescence in stratiform clouds. *Quart. J. Roy. Meteor. Soc.* **78**, 377–386.

MASON, B. J. (1971) *The Physics of Clouds.* Clarendon Press, Oxford, 671 pp.

MASON, B. J., and P. R. JONAS (1974) The evolution of droplet spectra and large droplets by condensation in cumulus clouds. *Quart. J. Roy. Meteor. Soc.* **100**, 23–38.

MAZIN, I. P. (1968) The stochastic condensation and its effect on the formation of cloud drop size distribution. *Proc. Intl. Conf. on Cloud Physics,* Toronto, pp. 67–71.

McDONALD, J. E. (1958) The physics of cloud modification. *Advances in Geophysics,* vol. 5, pp. 223–303, Academic Press Inc., New York.

McDONALD, J. E. (1963a) Early developments in the theory of the saturated adiabatic process. *Bull. Amer. Meteor. Soc.* **44**, 203–211.

McDONALD, J. E. (1963b) Use of the electrostatic analogy in studies of ice crystal growth. *Z. Angew. Math. Phys.* **14**, 610–620.

MELZAK, A. Z., and W. HITSCHFELD (1953) A mathematical treatment of random coalescence. McGill University, Stormy Weather Group Rep. MW-11, 28 pp.

MIDDLETON, W. E. K. (1966) *A History of the Theories of Rain.* Franklin Watts, Inc., New York, 223 pp.

MORDY, W. (1959) Computations of the growth by condensation of a population of cloud droplets. *Tellus* **11**, 16–44.

MURRAY, F. W., and L. R. KOENIG (1972) Numerical experiments on the relation between microphysics and dynamics in cumulus convection. *Mon. Wea. Rev.* **100**, 717–732.

NAKAYA, U. (1954) *Snow Crystals.* Harvard University Press, 521 pp.

NEWTON, C. W. (1967) Severe convective storms. *Advances in Geophysics,* vol. 12, pp. 257–308, Academic Press Inc., New York.

OGURA, Y., and T. TAKAHASHI (1971) Numerical simulation of the life cycle of a thunderstorm cell. *Mon. Wea. Rev.* **99**, 895–911.

PARUNGO, F. P., and J. P. LODGE (1965) Molecular structure and ice nucleation of some organics. *J. Atmos. Sci.* **22**, 309–313.

PITTER, R. L., and H. R. PRUPPACHER (1974) A numerical investigation of collision efficiencies of simple ice plates colliding with supercooled water drops. *J. Atmos. Sci.* **31**, 551–559.

REINHARDT, R. L. (1972) An analysis of improved numerical solutions to the stochastic collection equation for cloud droplets. Ph.D. thesis, University of Nevada, 111 pp.

RIGBY, E. C., J. S. MARSHALL, and W. HITSCHFELD (1954) The development of the size distribution of raindrops during their fall. *J. Meteor.* **11**, 362–372.

ROBERTSON, D. (1974) Monte Carlo simulations of drop growth by accretion. *J. Atmos. Sci.* **31**, 1344–1350.

ROGERS, R. R. (1967) Doppler radar investigation of Hawaiian rain. *Tellus* **19**, 432–455.

ROSINSKI, J. (1974) Role of aerosol particles in formation of precipitation. *Rev. Geophys. and Space Phys.* **12**, 129–134.

SCORER, R. S. (1958) *Natural Aerodynamics.* Pergamon Press, Oxford, 312 pp.

SCORER, R. S., and H. WEXLER (1963) *A Colour Guide to Clouds.* Pergamon Press, Oxford, 63 pp.

SCOTT, W. T. (1968) On the connection between the Telford and kinetic equation approaches to droplet coalescence theory. *J. Atmos. Sci.* **25**, 871–873.

SCOTT, W. T. (1972) Comments on "Validity of the finite-difference droplet collection equation". *J. Atmos. Sci.* **29**, 593–594.

SEKHON, R. S., and R. C. SRIVASTAVA (1970) Snow size spectra and radar reflectivity. *J. Atmos. Sci.* **27**, 299–307.

SLINN, W. G. N. (1975) Atmospheric aerosol particles in surface-level air. *Atmospheric Environment* **9**, 763–764.

SRIVASTAVA, R. C. (1967) A study of the effect of precipitation on cumulus dynamics. *J. Atmos. Sci.* **24**, 36–45.

SRIVASTAVA, R. C. (1971) Size distribution of raindrops generated by their breakup and coalescence. *J. Atmos. Sci.* **28**, 410–415.

STEINER, J. T. (1973) A three-dimensional model of cumulus cloud development. *J. Atmos. Sci.* **30**, 414–435.

TAKEDA, T. (1971) Numerical simulation of a precipitating convective cloud: the formation of a "long-lasting" cloud. *J. Atmos. Sci.* **28**, 350–376.

TELFORD, J. W. (1955) A new aspect of coalescence theory. *J. Meteor.* **12**, 436–444.

THOMSON, W. (1870) On the equilibrium of vapour at a curved surface of liquid. *Proc. Roy. Soc. Edinb.* **7**, 63–68.

TWOMEY, S. (1959) The nuclei of natural cloud formation: the supersaturation in natural clouds and the variation of cloud droplet concentration. *Geofis. Pura et Appl.* **43**, 243–249.

TWOMEY, S. (1964) Statistical effects in the evolution of a distribution of cloud droplets by coalescence. *J. Atmos. Sci.* **21**, 553–557.

TWOMEY, S. (1966) Computations of rain formation by coalescence. *J. Atmos. Sci.* **23**, 405–411.

VALI, G. (1968) Ice nucleation relevant to formation of hail. McGill University, Stormy Weather Group Rep. MW-58, 51 pp.

WARNER, J. (1969a) The microstructure of cumulus cloud. Part I. General features of the droplet spectrum. *J. Atmos. Sci.* **26**, 1049–1059.

WARNER, J. (1969b) The microstructure of cumulus cloud. Part II. The effect on droplet size distribution of the cloud nucleus spectrum and updraft velocity. *J. Atmos. Sci.* **26**, 1272–1282.

WARSHAW, M. (1967) Cloud droplet growth: statistical foundations and a one-dimensional sedimentation model. *J. Atmos. Sci.* **24**, 278–286.

WARSHAW, M. (1968) Cloud drop coalescence: effects of the Davis-Sartor collision efficiency. *J. Atmos. Sci.* **25**, 874–877.

WEXLER, R., and P. M. AUSTIN (1954) Radar signal intensity from different levels in steady snow. Massachusetts Institute of Technology, Weather Radar Research, Rep. No. 23, 27 pp.

WEXLER, R. (1955) An evaluation of the physical effects in the melting layer. *Proc. Fifth Wea. Radar Conf.*, Ft. Monmouth, N.J., pp. 329–334.

WEXLER, R. (1960) Efficiency of natural rain. *Physics of Precipitation* (H. Weickmann, ed.), Geophys. Monograph 5, American Geophysical Union, pp. 158–163.

YOUNG, K. C. (1974) The evolution of drop spectra through condensation, coalescence and breakup. *Preprints, Conference on Cloud Physics*, Tucson, Ariz., pp. 95–98.

YOUNG, K. C. (1975) The evolution of drop spectra due to condensation, coalescence and breakup. *J. Atmos. Sci.* **32**, 965–973.

ZAWADZKI, I. I. (1973) Statistical properties of precipitation patterns. *J. Appl. Meteor.* **12**, 459–472.

ANSWERS TO ODD-NUMBERED PROBLEMS

Chapter 1

1. Each constituent gas separately obeys the ideal gas law, $p_n = R^* M_n T / m_n V$, where p_n, M_n, and m_n denote, respectively, the pressure, mass, and molecular weight of the nth constituent. The volume occupied by the mixture is V. By Dalton's law, the total pressure is the sum of the partial pressures, $p = \Sigma p_n = (R^* T / V) \Sigma (M_n / m_n)$. Therefore $p\alpha = R^* T \Sigma (M_n / m_n) / \Sigma M_n$ and $\overline{m}$, the effective mean molecular weight of the mixture, is given by
$$\overline{m} = \Sigma M_n / \Sigma (M_n / m_n).$$

3. Final temperature = 198 K. Work done by the air = 258 J. Heat added to air = 216 cal.

5. The tephigram coordinates are $c_p T$ versus $\ln \theta$. From (1.24), along any isobar $\partial \ln \theta / \partial \ln T)_p = 1$. Therefore the slope of an isobar on a tephigram is given by $(\partial \ln \theta / \partial c_p T)_p = 1 / c_p T$.

Chapter 2

1. For water vapor undergoing an adiabatic expansion, $(de/dT)_{adiab} = c_{pv} / \alpha_v$. From (2.10), $de_s/dT = L/T\alpha_v$. Condensation can occur in adiabatic expansion if $(de/dT)_{adiab} < de_s/dT$, which implies $T < L/c_{pv}$. This inequality is satisfied.

3. The differential of w_s is given by $dw_s = \varepsilon p e_s (d \ln e_s - d \ln p) / (p - e_s)^2$. Employing the Clausius–Clapeyron equation for de_s/e_s and combining terms leads to $dT/dp = (T/p)(A/B)$ in the pseudoadiabatic process, with A and B positive. Thus dT/dp is positive and temperature decreases with decreasing pressure.

5. (a) 280.8 K; (b) 5.6 g/m³; (c) 72%; (d) 275.5 K; (e) 277.5 K; (f) 289 K; (g) 294 K; (h) 302 K.

Chapter 3

1. (a) $- \partial T / \partial z = g / R' = 3.41 \times 10^{-2}$ K/m.
(b) $H = p_0 / \rho g = R' T_0 / g$.

3. Employing the definition of geopotential and the hydrostatic equation shows that $d\psi = - R' T_v (dp/p)$ which, when integrated, gives the indicated result.

5. With subscripts 1 and 2 denoting conditions over land and water respectively, the top of the circulation is at a height of

$$H = \frac{(R'/g) \ln(p_2/p_1)}{(1/T_2) - (1/T_1)} = 980 \text{ m}.$$

Chapter 4

1. In general, $\dfrac{1}{f}\dfrac{\partial f}{\partial z} = \dfrac{1}{w}\dfrac{\partial w}{\partial z} + \dfrac{L\varepsilon\gamma}{R'T^2} - \dfrac{\Gamma c_p}{R'T}$.

When well mixed, $\dfrac{1}{f}\dfrac{\partial f}{\partial z} = \dfrac{\Gamma}{RT}\left(\dfrac{L\varepsilon}{T} - c_p\right) > 0$ because $T < L\varepsilon/c_p$.

For $T = 250$ K and $f = 50\%$, $\partial f/\partial z = 3.6\%$ per 100 m.

3. (a) 24.2 m/sec; (b) 24.4 m/sec; (c) 22.9 m/sec.

5. p.w. $= \displaystyle\int_0^H w\rho\,dz = \dfrac{1}{g}\int_{p_H}^{p_0} w\,dp.$

For specific example, p.w. $= 2.04$ cm.

7. $\alpha = -1/4$; $\beta = -5/4$.

Chapter 5

1. Employing (5.1) and (2.10), we find that

$$\frac{\partial \ln e_s(r)}{\partial T} = \left(\frac{L}{R_v} - \frac{2B}{rR_v\rho_L}\right)\frac{1}{T^2} > 0 \quad \text{if} \quad r > \frac{2B}{L\rho_L}$$

When $r = 0.13\ \mu$m the change in $e_s(r)$ per degree of temperature increase is approximately equal to the change in $e_s(r)$ per micron of radius decrease.

Chapter 6

1. Comparing the heat balance equation (6.10) with the equation for the wet-bulb temperature (2.23) shows that the drop temperature and the wet-bulb temperature are related by

$$(T_r - T)/(T_w - T) = D\rho c_p/K.$$

From the data in Table 6.1, this ratio is found to equal approximately 1.19, implying that $T_r \approx T_w$. Mason (1971) has suggested that the accepted values of D may be about 10% too high. If so, then $T_r \approx T_w$ to a closer approximation.

Chapter 7

1. Final diameter $= 1.70$ mm.

3. (a) 0.35 mm radius; (b) 20.9 min; (c) 0.38 mm radius.

5. 280 m.

Chapter 8

1. $T_r = -21.5°$C.

3. The heat balance relation (8.2) is satisfied if $T_r \approx -18.3°C$, corresponding to $\rho_{vr} = 1.04$ g/m³. Solving (8.1) then shows that the crystal must have an initial radius of 6.5 cm to survive the fall, which is unreasonably large.

Chapter 9

1. Time required is 5.58 min, during which the particle ascends 3.10 km.

3. Change in cloud water content due to sweepout is given by

$$\left(\frac{dM}{dt}\right)_{acc} = -\frac{3}{2}\pi E N_o k M R^{4/5} (4\pi N \, k)^{-4/5}.$$

Condensation rate is given by

$$\left(\frac{dM}{dt}\right)_{cond} = \rho^2 U \Gamma_s \varepsilon L w_s / p T$$

Balance exists when

$$\left(\frac{dM}{dt}\right)_{acc} + \left(\frac{dM}{dt}\right)_{cond} = 0.$$

For typical conditions the balance implies, numerically and in c.g.s. units,
$$(U/M)R^{-4/5} = 9 \times 10^{10} E.$$
Note that the condensation rate, following from (6.22), may also be obtained from

$$\left(\frac{dM}{dt}\right)_{cond} = \rho \frac{d\chi}{dt} = \rho \frac{Q_1}{Q_2} U,$$

which is equivalent to the expression given earlier.

Chapter 10

1. $Z = 6!(4\pi k)^{-7/5} N_0^{-2/5} R^{7/5}.$
   For Z in mm⁶/m³ and R in mm/hr, $Z = 216 R^{1.4}$.

3. At time t the reflectivity factor is given by

$$Z(t) = 2^6 \int_0^\infty n(r,t) r^6 dr.$$

Droplet growth is described by $r(t) = (r_0^2 + \alpha)^{1/2}$, where $\alpha = 2t(S-1)/(F_k + F_d)$. In terms of the initial droplet spectrum, $n(r,t) = n_0(r_0)(dr_0/dr)$. Therefore

$$Z(t) = 2^6 \int_0^\infty n_0(r_0)(r_0^2 + \alpha)^3 dr_0$$

$$= 2^6 N[\overline{r^6} + 3\alpha\overline{r^4} + 3\alpha^2\overline{r^2} + \alpha^3],$$

where N is the droplet concentration and $\overline{r^j}$ is the mean jth power of radius in the initial distribution. Since $Z(0) = 2^6 N \overline{r^6}$, it follows that

$$\frac{Z(t)}{Z(0)} = 1 + (3\alpha \overline{r^4} + 3\alpha^2 \overline{r^2} + \alpha^3)/\overline{r^6}. \tag{A}$$

The even moments of a Gaussian distribution are

$$\overline{r^2} = \sigma^2 + (\overline{r})^2; \quad \overline{r^4} = 3\sigma^4 + 6\sigma^2(\overline{r})^2 + (\overline{r})^4;$$

$$\overline{r^6} = 15\sigma^6 + 45\sigma^4(\overline{r})^2 + 15\sigma^2(\overline{r})^4 + (\overline{r})^6.$$

Substituting these into (A) and employing $\sigma = 0.15r$ gives

$$\frac{Z(t)}{Z(0)} = 1 + (3.41\alpha(\overline{r})^4 + 3.07\alpha^2(\overline{r})^2 + \alpha^3)/(\overline{r})^6. \tag{B}$$

For the conditions given, $F_k + F_d \approx 0.14 \times 10^7$ sec/cm².

Consequently, the reflectivity at time t, measured in decibels above the initial value, is

$$10 \log \frac{Z(t)}{Z(0)} = 19.3 \text{ dB at 5 min}$$

$$= 26.6 \text{ dB at 10 min}$$

$$= 31.3 \text{ dB at 15 min.}$$

Notice that for a monodisperse distribution ($\sigma \to 0$),

$$\frac{Z(t)}{Z(0)} = 1 + (3\alpha r^4 + 3\alpha^2 r^2 + \alpha^3)/r^6,$$

which is a very good approximation to the more accurate result (B).

Chapter 11

1. So defined, the sweepout efficiency of a particle equals the product of its velocity times its cross-section. Thus, $E(\text{graupel})/E(\text{rain}) = 0.48 \ m^{-2/15}$ where m is the particle mass. It can be further shown that graupel has a higher sweepout efficiency than a raindrop of equal mass if the graupel radius $r_g < 0.2$ cm. Conversely, if $r_g > 0.2$ cm the raindrop has the higher efficiency.

3. The reflectivity factor at time t is given by

$$Z(t) = 2^6 \int_0^\infty n(r,t) r^6 dr.$$

But the drop-size distribution evolves according to (11.3). Therefore

$$Z(t) = 2^6 \int_0^\infty r^6 e^{-at} n_0 (re^{-at}) dr$$

which reduces to

$$Z(t) = e^{6at}Z(0).$$

The signal increases by 10 dB when

$$t = \ln 10/6a \approx 77 \text{ sec.}$$

5. Rain rate is given by

$$R = \frac{\pi}{6} \int_0^\infty N(D)D^3V(D)dD = \pi N_0 k 4!/6b^5.$$

Rate of change of cloud water due to rain-collection is

$$\frac{dM}{dt} = -\frac{\pi}{4} \int_0^\infty D^2 N(D)V(D)EMdD.$$

Therefore

$$\frac{dM}{M} = -\frac{3}{2}E(\pi k N_0)^{1/5} 4^{-4/5} R^{4/5}dt.$$

For $E \approx 1$ and R expressed in consistent units, this leads to the required result.

Chapter 12

1. The heat balance implies that

$$REML_fU(R) = 4K(T_s - T_0 - \gamma h)b$$

where γ is the lapse rate and h is the distance of fall from 5 km, where the ambient temperature is T_0. The ventilation factor may be shown to be of the form $b = k_2 R^{3/4}$, where $k_2 \approx 50$. The growth rate is given by

$$dR/dh = EM/4\rho_s.$$

Solving for EM from the balance relation, substituting into the growth equation and integrating, leads to the indicated result.

3. The rate of change of temperature of the hailstone is given by

$$dT_s/dt = -(mc_s)^{-1} dQ_s/dt,$$

where m and c_s denote its mass and specific heat. From (12.3) this equation may be written

$$dT_s/dt + \alpha T_s = \beta$$

where $\alpha = 4\pi RKb/mc_s$ and $\beta = \alpha T$.

The solution of this equation is

$$(T_s - T) = (T_0 - T)e^{-\alpha t},$$

where $T_0 = T_s(0)$. Therefore the time constant τ is given by $\tau = 1/\alpha = R^2\rho_s c_s/3Kb$. For the conditions specified the time constant amounts to approximately 26 sec.

Chapter 13

1. The depletion of cloud water by accretion is given by

$$\left(\frac{dM}{dt}\right)_{acc} = -\pi E M D^2 N V(D)/4.$$

The condensation rate may be determined in various ways (see solution to problem 3, Chapter 9) and is given approximately by

$$\frac{1}{U}\left(\frac{dM}{dt}\right)_{cond} = 5 \times 10^{-12} \qquad \text{(c.g.s. units)}$$

Under the conditions given, the balance requires that the concentration N of small hailstones must exceed approximately 100 per m^3, or 0.1 per liter.

INDEX

OTHER TITLES IN THE SERIES IN NATURAL PHILOSOPHY